# Learning To Manage Technical Professionals

## Crossing the Swamp

# Learning To Manage Technical Professionals

## Crossing the Swamp

Richard J. Stein

ADDISON-WESLEY PUBLISHING COMPANY
Reading, Massachusetts • Menlo Park, California • New York
Don Mills, Ontario • Wokingham, England • Amsterdam
Bonn • Paris • Milan • Madrid • Sydney • Singapore
Tokyo • Taipei • Mexico City • San Juan

The publisher offers discounts on this book when ordered in quantity for special sales. For more information, please contact:

    Corporate & Professional Publishing Group
    Addison-Wesley Publishing Company
    One Jacob Way
    Reading, Massachusetts 01867

**Library of Congress Cataloging-in-Publication Data**
Stein, Richard J.
  Learning to manage technical professionals : crossing the swamp / Richard J. Stein
    p.  cm.
  "Published simultaneously in Canada." —CIP galley.
  Includes bibliographical references.
  ISBN 0-201-63320-5 (alk. paper)
  1. High technology industries—Management.  2. High technology industries—Personnel management.  I. Title.
HD62.37.S74   1993
620'.0068—dc20                                                            93-3014
                                                                                                          CIP

Copyright © 1993 by Addison-Wesley Publishing Company
All rights reserved. No part of this publication may be reproduced, stored in a retrieval system, or transmitted, in any form, or by any means, electronic, mechanical, photocopying, recording, or otherwise, without the prior consent of the publisher. Printed in the United States of America. Published simultaneously in Canada.

Text design by Wilson Graphics & Design (Kenneth J. Wilson)
Set in 10-point Palatino by Gex Publishing Services

ISBN 0–201–63320–5

Text printed on recycled and acid-free paper.
1 2 3 4 5 6 7 8 9 10 – MU – 96959493
First Printing, June 1993

# Contents

**Introduction to the Swamp**  ix

Chapter 1  **You're Promoted — Now What?**  1
    Survival in the First Weeks  1
    Day One: Define the Job; Establish Control  2
    Week One: Test the Controls  4
    Week Two and Beyond: Plans, Paper, People  6
    The Managerial Skills: Delegation, Communication, Direction  8
    Alligators  10
    Summary  12

Chapter 2  **The High-tech Difference**  13
    Continuous Innovation  14
    Responding to Market Changes  17
    Sensitivity to Small Technical Decisions  19
    Capability for Rapid Reorganization  19
    An Employer of Highly Skilled Specialists  20
    Knowledge Hunger  21
    Lean in Structure, Fat in Technical Smarts  21
    High-tech Management Characteristics  22
    The High-tech Manager  24
    The Management Zoo  27
    Motivation  30
    Resilience  31
    Your Job  31

Chapter 3  **Tiger Team, Skunkworks, or What?**  35
    Culture  35
    The Organizational Structure  37

How to Do It  39
Your Tiger Team  41
Who's on the Team?  42
Match Skills with Tasks  43
Turning Over the Rocks  45
Which Rocks to Turn Over  45
Interview: Art or Fraud?  48
The New Challenge of Diversity  50
Alligators  54

Chapter 4  **Planning Innovation**  55

Planning  55
Here to There  56
Details  57
The Backwardness of Plans — The Objective  58
Strategy and Tactics  61
Milestones  63
Strategic Planning  64
Resources  64
Making a Good Plan  68
Critical Elements of Any Plan  69
Planning in the Innovation Business  72
Setting the Stage  73
On the Plan  74
Whose Objective? A Case Study  76

Chapter 5  **Working Faster**  77

Speeding Up  77
Product Cycle Time  78
The Common Agenda  80
Competition and Acceleration  81
Proof of Concept  82
Evaluating Manufacturability  82
Bottlenecks  83
Dead Ends  84

　　　　　　　Making Your Outfit More Responsive   85
　　　　　　　How to Speed Up — A Summary   87

Chapter 6   **How to Create an Exceptional Team**   89
　　　　　　　The Tribe   89
　　　　　　　What Is a Group?   90
　　　　　　　A Group Organizes Itself   92
　　　　　　　Another Analogy   93
　　　　　　　Factions   94
　　　　　　　The Boundary   96
　　　　　　　Examine Thy Navel   97
　　　　　　　Motivation   98
　　　　　　　Delegation   100
　　　　　　　Porter's Complaint   103
　　　　　　　Delegation and the Tiger Team   104
　　　　　　　Making Meetings Work   105
　　　　　　　Communication   108
　　　　　　　Communicable Disease: Case Study   109
　　　　　　　Ron, Alone   110

Chapter 7   **Making Progress**   111
　　　　　　　How Do You Know What's Happening?   111
　　　　　　　Your Tailors   112
　　　　　　　Stealth Project   113
　　　　　　　Leaving Baggage Behind   113
　　　　　　　The NIH Syndrome   114
　　　　　　　Retroengineering   115
　　　　　　　Not to Hurt, Not to Kill   116
　　　　　　　How to Tell a Tough Job from an Easy One   117
　　　　　　　Is Your Group Good, Better, Best?   118
　　　　　　　Grading Your People   119
　　　　　　　Salaries and Other Rewards   123

Chapter 8   **Paying Your Way — Every Day**   127
　　　　　　　Doing the Manager Thing   127
　　　　　　　Measure Yourself   127
　　　　　　　Refine Your Plan   128

Fighting Fires   130
Pleasing Your Boss   131
Bite Your Tongue: When the Boss Says "Lie"   132
Leverage   133
Keep Learning   134
Advance Yourself   136
The Harvard Hula Hoop   137
An Alligator in Charge   138
Lazy Paul   139

Chapter 9   **Quality — Buzzword Or Religion?**   141

Quality Quiddity   141
Quality, Innovation, and High Tech   142
Where Did the Quality Revolution Come From?   142
Going After Quality   144
Using Quality Tools   146
How to Start a Quality-Improvement Program   146
Quality as a Company Goal   147
Quality Is a Cooperative Effort   149
Quality and Profit   150
Measuring Quality   150
Six Sygma Is No Misteak   152

Chapter 10   **Checklists**   153

Chapter 1   153
Chapter 2   154
Chapter 3   156
Chapter 4   157
Chapter 5   159
Chapter 6   160
Chapter 7   163
Chapter 8   164
Chapter 9   166

**Suggested Books and Periodicals**   169

# Introduction to the Swamp

You are a technical worker, a scientist, or an engineer. You have recently become a manager of a group of five to ten people. You came to this new job with no training in management, and possibly no interest. In fact, if you're like most of us, you don't think management is worth much study. You've been managed for a while, you know there are more paperwork and more meetings than you'd like, and you assume that the rest of it is either obvious or inconsequential.

You're also in a high-tech business, one which depends on innovation, rapid change, and very skilled people. The reading you've done on management techniques isn't too helpful . . . your job is different. All the conventional material is either written for company presidents and concerns major policy directions and high-level strategy, or else it appears to be useful on the (extinct kind of) factory floor.

Worse, there are "non-technical" components in management, like interpersonal relations, motivation, group dynamics, and the effect of paint color on mood in offices. These are either irrelevant or excessively mushy subjects, the way you see it. Your people are all self-motivated, they know what they have to do, and all the problems should be quantifiable.

Besides, how can anyone do serious planning when the whole world turns over every year, competition is fierce, employees come and go with the seasons, and your product doesn't look like anything else ever made? How can you increase the quality of a research program? What's the standard deviation of innovation?

This is the high-tech-management swamp. It is a place where the footing is slippery, the water murky, and alligators plentiful. It is a far different place from the clear and specific world of ordinary engineering, where you used to walk. Instead of black-and-white facts and figures, you now have to deal with uncertainty about what is possible and how to plan for it. Instead of doing some work at the bench, you have to figure out how to get *someone else* to do the work. Instead of just worrying about technical feasibility, you have to worry about cost, saleability, the market, and what your boss really wants.

I use the alligator throughout this book as a symbol of problem areas and dangerous situations that can suddenly give you a nasty nip or even end your career. The nice aspect of alligators, though, is that they are not completely invisible. In fact, once you know what to look for and where to look, you can see them coming and fend them off. A new manager doesn't always attach significance to those two little eyes breaking the surface of the water and might get bitten as a consequence. If this book alerts you to even one alligator, it should be worth your time.

I also like the idea of a swamp. One of a manager's primary jobs is planning, which is really navigating from some here-and-now wish to some then-and-there result. Navigation in a swamp is complicated by the lack of distinct landmarks, the impossibility of direct routes, and difficulty in taking fixes on the stars. You will never have enough information to "master" this environment, but you can adapt to living in it.

This book doesn't have all the answers. It's an attempt to help people like you, new managers in high tech, who are faced with unique and difficult problems every day. It assumes that your company isn't going to send you to school for a couple of years to get credentials in management. A weekend seminar in company team-building might be considered a luxury. You'll be lucky to get away for a few afternoons of watching videotapes on the subject!

What I've tried to do is touch on a lot of subjects — planning, organization, motivation, quality, hiring, and career growth — which have been covered in countless other books. What makes this book different is that it focuses on how management problems differ in technology. I've tried to emphasize problems that are especially difficult to solve in our type of business. Sometimes the methods and solutions are exactly the same as they are in other kinds of business, and sometimes they differ. Sometimes the suggestions are radical and can't be used in your company. Your company culture may interfere, resources may not be available, or you may have a completely unique problem.

Some of the techniques I describe are obvious, elementary, and already used by anyone except a pure novice. Some are advanced in the sense that you'd have to have some experience to recognize or cope with the associated problems. No book can substitute for working experience or the advice of mentors and experts. I've borrowed an icon from the ski slopes to help you identify the tricky stuff: a black diamond (♦) indicates expert terrain.

I've been a participant in R&D, in both government organizations (the National Bureau of Standards) and commercial enterprises (Texas Instruments, Balzers, Perkin-Elmer), for long enough to have noticed that there are thousands of us who need specialized management materials but who are too jammed up with work to do any serious study. I've also noticed that people like us have been ignored in the management literature. I've seen many good ideas, good people, and good companies crash because of bad management. One of the differences in high tech is that grunt-level technical and management errors can have tremendous consequences. The actions of everyone, from the junior engineer specifying a component to the research guru coming up with breakthrough concepts, can determine the success or failure of the company.

Meanwhile, the United States is fighting a global battle in the marketplace. Our overseas competitors do not manage the way we do. They invest much more in training, and they outspend us in research, manufacturing technology, and product development. We are losing in nearly all the leading-edge areas, and we are not doing much about it. You may be awash in buzzwords and poster-engineering about quality and new (old Japanese) manufacturing programs, but it isn't necessarily doing you much good. You have new tools, such as computer-based modeling and design, to speed up your job, but you will discover that just as many old management practices are working to slow it down.

You can't fix these big problems, but you can contribute where it counts the most, at the bench, in the lab, where the technology is created and made real. If you save a few hours by running better meetings or eliminating memos, you might be able to design a better product. If you figure out how to talk to and motivate that nearly invisible loner engineer, you may uncork an effort that puts your company ahead. If you can write up a plan that allows for dead ends and breakthroughs, you can avoid burnout-provoking whiplash changes of direction. You may even get more and better product out the door.

All these skills are part of effective management, and they're available to be learned. These skills are not trivial or inconsequential. Your job has changed — and you have to change, too.

# Chapter 1

# You're Promoted — Now What?

### SURVIVAL IN THE FIRST WEEKS

Here it is. You're no longer an engineer, a scientist, or a technician — you're a *manager*! The idea of being a manager sounds good, and you probably have a few ideas about how you're going to help the company make better decisions, how you're going to cut out some of the administrative flypaper, how your group will be the best, and how everyone will be rewarded with honors and coins. You're in charge!

But —

You have absolutely no background in management. You've always thought of it as a vague, soft, nonchallenging pursuit taken up by people who are not like you and me, either because they never were good at technical work, or because they somehow lost interest. You also probably have a picture of administrative work as dull, unproductive paper-pushing mixed with boring, pointless meetings.

If you're like most technical sorts, you want to remain a working technologist and not become "obsolete." You also don't feel really confident about flailing around in a field you don't know much about. If you've been an engineer, you want to remain an engineer. You want to do the things you're good at, tasks you feel are important and rewarding. *What you do want is to remain a maker and implementer of technical decisions — a creator of ideas and things.*

Frankly, you're not sure that management is *that* important, and other than cash, you're not convinced that the rewards are as clear as the kick you get from seeing your concepts fleshed out in products. The truth of the matter is that management is not only challenging, but it is critically important to success in high tech. Hopefully, you will come to understand why this is so, and why you, as a manager, can remain a creator of ideas and things.

While you're considering the deeper implications of your new role, don't come to a complete halt, because the first few weeks of your new job are critical. Any mistakes you make now, especially in interpersonal areas, will propagate and give you unnecessary grief later on. You need to hit the ground running and learn enough to get your new organization moving in the right direction.

## DAY ONE: DEFINE THE JOB; ESTABLISH CONTROL

The only reason you've been made a manager is to achieve a specific company objective. What is it? It may be to create a component of a new product or process. It may be making an existing product better, faster, or cheaper. It may even be helping to sell something. Any outfit worth its salt has no open-ended objective like "to do engineering" or "to do research." Even the leading-edge powerhouses of our trade, such as (what was) Bell Labs or IBM Watson, nurture long-range work only if there is a potential connection with future business.

Presumably your boss defined your new job, or you think you know what it is. However, get it down on paper right now, and then answer these questions:

- Is the objective specific enough?
- Is there a plan, or do I make one?
- What resources do I need to do this job?
- How long do I have to do it?

Once you have a solid feel for the objective (which may change tomorrow or never), you have to confirm your job by establishing control.

The importance of taking control quickly can't be overstressed. The first impression you make as a manager is likely the only one you get to make, at least with your own people. The faster you demonstrate that you *can* lead the group, the less damage is done by your assignment.

Damage? Lots of inexperienced managers, especially if they've been promoted during an existing project, assume that work progresses smoothly during a change of authority. It doesn't. Everyone, including those self-motivated scientists, gets uneasy and worried if the new boss isn't talking and listening (especially listening) from day one. What has actually happened in the transition that put you in charge is that the whole invisible group organization, pecking order, dynamic, or whatever, has been turned to dust. The people look the same, the desks and machines look the same, but all the connections are ripped. To make a punk analogy, it's like a bunch of chemical bonds that have been broken: they're waving around in space, looking for places to reattach. Meanwhile, what was a compound, or an organization, is next to useless. Unless you dive in and direct the reaction, some of those bonds are going to be made in the wrong places.

My most effective bosses gave the initial impression that they knew the project, the customer, the tools, and techniques. They were able to make decisions quickly, while not appearing to be too autocratic. They made sure, the first day, that the whole group was moving in a defined and reasonable direction.

My worst boss, on the other hand, was such an ephemeral character that we were never certain that we reported to him at all. We got our orders from another man, went to him with questions, and we felt that we *worked* for the other guy!

How do you establish control? First, you have to meet everybody who will report to you, both as a group and individually. Get three concepts across:

1. You are the focus of all questions and authorizations.
2. The group objective is _____.
3. You want everyone's opinion on the objective.

Exactly how you continue from here depends on whether your job is research, manufacturing, or whatever. You may have one big meeting, or you may introduce yourself one-on-one. You could provoke fear with threats, admiration with bragging, or loyalty with promises. There is such a thing as a management style, and you will develop one . . . eventually. For now, what method you should use is not certain until you figure out how your group will work best. It's difficult to say how much you emphasize your control versus group decision-making, because there are real and significant differences in how small groups function. Unless you already know everyone and how they relate to one another, the first day is too soon to figure this out. Mostly, it will take weeks or months of studying your group and the way it works. The idea in all cases, however, is to somehow show that there is a job, a group, and a boss.

There are a few distinct things you should *not* do:

- Don't say everything continues as before. Your group is in an uncertain state the first day, waiting for orders. The tigers are expecting positive change, new and better goals, and more recognition for their efforts. Even the below-average players are hoping for some change.

- Don't pass the buck to your boss or other bosses. You should be trying hard to be recognized as being in control of the group. When you defer decisions to others, you drop out of the control loop.

- Don't claim to know everything. You're working with an intelligent crew, each of whom knows more than you do about many subjects, possibly including company strategic objectives, administrative procedures, and other managerial things. In fact, it is likely your willingness to admit ignorance which will get you *elected* leader.

- Don't put projects or people on hold. Momentum is important. When you stop work to get a fix on where you are, perhaps by asking for progress reports, summaries, or evaluations, you're sending a clear message that you do not know which way to go, or that you cannot make decisions. Brainwork has considerable momentum. It is therefore difficult to get started and difficult to stop, or even redirect. You need to keep the constructive momentum in the early, critical period, when you get on board.

Even if you know that major changes are in store, that the staff may be added to or cut back, that some work will be discontinued, you need to avoid first-day mishandling of that uncertain and sensitive group.

---

**What Do You Know After a Day on the Job, If You're Lucky?**

- The objective — why you're there
- Who reports to you directly and indirectly
- Everybody's face and function
- You're the focus of authority and communication

---

## WEEK ONE: TEST THE CONTROLS

Management is a trip from *here* to *there*. To get *there* you need to have a plan, and controlling and directing people is how you travel. If you intend to fly a plane, you first walk around the outside, look over all the control surfaces, kick the tires, uncover the airspeed tube, look at the oil. You climb in, push the brakes, test the rudder pedals, move the yoke all around, run up the motor, check the dials, and in fact, go line by line through a checklist even if you're certain everything works.

*Control?* Isn't that a dirty word? Don't we say *supervise*? 'Fraid not. You don't "supervise" that plane, your computer, or your car. You control them. Managers control and direct people. You achieve objectives by actively causing work to be performed. It is not a passive occupation.

Before you risk company money, your career, and other assets, you'd better test all the controls to see what they all do and whether they work right. Don't assume anything. Test. Can the secretary answer the phone? Can the engineer read a print? Can a researcher research? Can the supporting services deliver their product? Does the physical equipment work? Is it calibrated?

During the first week, while you're being buried under the snowstorm of idiot paperwork and trying to get in mesh with a mass of alien reporting requirements, test the controls. Don't make up phony work; just use what comes along. Write a letter and see how long it takes to clear the door. See if a fax from outside gets to your desk the same day it's sent. Order some supplies. Have something quick and simple designed or built. Try to personally give everyone in the group some short, unambiguous task. You will get some of this information from meetings, but technical professionals are notoriously taciturn in public, and it's not likely that anyone will flaunt his or her particular ignorance in front of a group.

By the end of the first week, your goal is to have made enough contact with your group to convince everyone that you are actually present and that you are functioning as a manager. You have established a primary *communication* channel from them to you, have *delegated* a few tasks, and have met with all parties to listen to their ideas on the major *objective*.

What you haven't done so far, is your primary job as a manager. You have not yet constructed a *plan*. That's okay. You can't do everything at once, and the first few days are committed to defining and staking out territory. (Yes, it *is* territory, it *is* yours, and you are *in control*.) You have also learned about your group, built some credibility, and kept the work flowing.

### Alligators

Here are some alligators to watch for:

- You're not starting out in equilibrium. The promotion is usually the result of external forces, such as a reorganization in the company, your ex-boss leaving suddenly, the advent of a new project, or a change in the marketplace. High tech is very much controlled by quick change elsewhere. Things happen quickly, and you're expected to become productive equally quickly. You have control over an organization that is in the middle of some kind of a transition — a nonequilibrium state.

- Worse, you've been promoted because some project has been going wrong, and your job is to fix it. Again, time is important, and you're tempted to cut corners in order to show results. Don't.

- For some reason, your group will not accept you as its leader. There's a big difference between working *with* a person and working *for* that person. Former peers, your colleagues, may not like the idea of you as their manager. At the worst, they will dig in their heels, complain, leave, or undercut you. This matters more than you think and is characteristic of any business that uses smart, self-directed people, as high-tech companies do.

- Your boss wants detailed planning before you get enough information to start the process. Planning is not a casual activity, especially if it's new to you. Hold off until you know what you've got to do, what the resources are, and how well the organization can follow.

- You have been given a detailed plan to work from, and it's obviously fantasy. The time scale is too short, too many breakthroughs are needed, or the concepts are just bad. You knew this going in, but the promise of a promotion was too tempting, and you thought that you'd magically escape disaster. Magical solutions belong to people with lots of experience. You're going to have to stick to that unworkable plan until you can establish enough credibility to change it.

---

**What Do You Know After a Week on the Job, If You're Lucky?**

- How the group works or doesn't work
- Which individuals can handle their jobs
- Whether the existing plan is off-base
- That there's a lot of administrative stuff to learn
- Some opinions on the objective

---

## WEEK TWO AND BEYOND: PLANS, PAPER, PEOPLE

The list of questions you need to answer increases to include the following quantifiables:

- What's the objective?
- Is there a plan?
- What tools (resources) are needed?
- Who are the players?

and these conceptual ones:

- What does everybody do?
- How do I know whether they're doing it?
- How do I direct and control?
- How do I delegate?
- How do I communicate?
- Where am I in the company?

These questions will be covered in detail in later chapters. For the time being, here are some pointers:

### Objectives

When you buy into a managerial job, you're buying into a serious contract between you and your company to deliver a product — a management product. You and your boss had better be perfectly aligned on the objectives or deliverables. "To do Joe's old job" is not an adequate objective! As soon as the dust settles a little, go back to your boss specifically to refine your joint notions of precisely what you're going to deliver. Have some ideas of your own, but don't press them too hard. Never act overwhelmed by the task. A manager is a person in control. Show that you're gathering vital information.

### The Plan

Making sense of the pre-existing plan is now a high-priority item. Your boss should be able to show you something in writing, with *objectives, milestones, dates, resources*. If the plan currently exists at your level in enough detail to work from, don't blindly accept it. Work through it carefully. Discuss the details with your boss. If you have suggestions, questions, alternative approaches, now is the time to talk. Both of you have to feel that you understand what the plan is, what motivated it, and where it can be changed.

♦ If a solid plan doesn't exist at your level, yours will still have to mesh with the plan your boss has — that is, his or her minor objectives become your primary ones, and the dates and budgets line up. Don't be afraid to do a little shooting from the hip; managers of managers like to see a display of quick analysis. Don't lock yourself in before you're relatively sure that you can do the job.

### Resources

Your resources are people, money, equipment, and plant. Chances are pretty good that you're familiar with the type of work that will be needed to satisfy the objectives. Chances are also good that you've never formally connected resources with a plan before now.

♦ Accounting concepts, in particular, are opaque to nearly everyone and have to be interpreted. Does a $50,000 employee cost a company $50,000, $245,000 or $0? Is a piece of equipment capital? Where's depreciation? And so forth.

♦ Understanding the present and potential capabilities of your staff is a hard job in itself. Quantifying their output can be a nightmare. How many innovations per staffer per year can you expect? You need two resource lists: what you've got, and what you need. Obviously, you need a pretty good first-cut plan to do this. Make your best guesstimates of resource requirements just to see whether you have to do anything immediate and drastic, such as adding a lot of staff. Remember that time constants are connected with any resource changes, and sometimes these time constants are longer than the project duration. I've seen more than one semiconductor company dither around for so long specifying, purchasing, and installing some flashy bit of tooling — an X-ray lithography system, to give a bad example — that the project window closes before the start button ever gets pushed.

### The Players

Unless you know otherwise, assume that you, as a new manager, have been given a fair share of real losers whom nobody else wants. Maybe your boss hopes that you have some unique touch that motivates the unmotivatable, or maybe your stamp is needed to finally eject some inert body. The sooner you find out who's good and who's not, the sooner you can do something about it. Watch out for uncontrollable *prima donnas*, people with excellent but irrelevant skills, and transients. A *prima donna* is the expert who believes that his or her rare skills merit exceptional attention. A transient is someone who never becomes part of your group.

If your star performer, having done a brilliant job, using big brushstrokes, decides that the details are trivial and boring, watch out! When anyone goes from workaholic to merely driven, watch out! When the test results aren't grabbed, hot and smoking, from the bench, watch out!

Try to keep your "bossness" from separating you from the group. Eat, drink, and play with them if you can.

## THE MANAGERIAL SKILLS: DELEGATION, COMMUNICATION, DIRECTION

Okay, here it is: The Manager Defined. A manager manages *people*. Surprise... you thought that objects, ideas, products, and paperwork were involved. They are, but these are not what's managed. You control and direct your staff, and they deal with the product. ♦♦ This distinction is hardly ever understood by technologists. In fact, quite a few of us don't even like the concept, because it puts us into that "no-longer-technical" category, or because it sounds too harsh. Get used to it ... it's what you do from now on. You:

- Define other people's tasks
- Provide resources and tools
- Plan and measure progress
- Represent, encourage, and interface for your group
- Help or hinder other people's careers
- Multiply your own capabilities

The skills you need to do the managerial job are considerably different from those in other professions. Just because there's no consistency in management curricula, the way there may be in mechanical engineering, perhaps, is no reason to write off the complexity or seriousness of the job. In fact, it may well turn out that you're a great engineer but can't manage worth a damn because you have trouble with people. Don't give up! The next decade will likely see the demise of the pure manager — that is, one without some detailed technical competence in his or her particular industry — and the rise of the technologist-manager. You're just going to have to work at it to stay in this crowd, but everyone else will, too.

So how does this translate to your second week on the job? By this time you have collected some information on objectives and resources. Now you have to start finding, and using some important management tools.

### The Manager's Toolbox

A good manager's toolbox contains the following:

- Planning, forecasting, and scheduling tools
- Resource acquisition tools
- Information tools
- Delegation, communication, and direction tools

*Planning tools* include methods and formalisms that assist in defining, visualizing, and measuring complex sequences of tasks. Each company, and frequently each division of a company, has a preferred way of doing planning, most of it expressed through software. If you don't know what's in use at your place, now is a good time to find out. The details don't matter nearly as much as understanding why you're doing it and how reliable or quantifiable it is. Read chapter 4.

*Resource acquisition tools* include methods for providing your staff with the space, time, money, and facilities needed to do the job. Again, your company may have several different methods for resource acquisition or transfer, and the more you know about them, the more likely it is that you can get what you need.

*Information tools* include those which help you relate your group's work with that of other groups, outside the company, in the technical literature, and in universities. These tools also help you interpret trade and business information that has an impact on your work. In high tech, especially, proper use of knowledge tools is critical throughout the organization. In fact, use of these tools is one of the best ways of identifying high tech. For example, these tools are being used when reports are well-referenced, the company library is busy, and cooperatives and consortia are being used.

*Delegation* is related to the concept of *leverage*. You, as a manager, have to multiply your own intellectual efforts through controlling the efforts of others. In other words, the amount of useful work (product) has to be increased by your presence more than enough to pay for your administrative costs. A manager who can't delegate is a manager who can be eliminated! Every time you dive into a nasty technical problem and temporarily stop doing the management job, you lose ground and stop paying for yourself. Many of us find it difficult to delegate, especially when we believe that we can do a certain job better. Beware — it's an alligator.

*Communication* is the most obvious characteristic of a manager. Written and verbal communication flow is continuous and often at elevated levels. A manager can be sandwiched between phone calls, meetings, reports, summaries, and presentations. Each company favors a different mix. Computer-related companies have always been heavy on e-mail. Japanese companies have depended on the telephone more than on paper. It seems to be a rule that as companies increase in size, the volume of internal communication increases. As we all know, at some point, the useful information carried by these channels is buried by the noise. ♦ Therefore, you have to be selective about communication, and you must learn skills — for example, in managing meetings — that can save you from becoming an uncritical shuffler of paper.

*A control tool* is one that causes someone else to do what you want done. In the military, these are called orders. In business, orders are replaced by requests and agreements. When you're dealing with specialized, skilled, creative, and intelligent people, direction is more a matter of *facilitation* than anything else: make it possible for someone to do the job, and they might just do it. Order it done, and you'll get garbage. You have to dip into the library of motivation and leadership here. ♦ The high-tech difference is that some of us have unusual motivators . . . ones never before described! The books I've seen fail to describe exactly why some people spend every waking hour on the job, living on fast food from vending machines, going home once every few days to change clothes, all to complete some bit of work that is of no interest to anyone else and that may or may not be useful to the employer.

## ALLIGATORS

Here are some alligators to watch for:

- *Administrative hogwash.* The paper blizzard won't bury you if you can stay ahead of it. Because you can't possibly know all the rules when starting out, you need help. The three people you should rely on are your boss, your secretary or assistant, and a financial person, such as a contract administrator. They will be more than happy to help you, because they know that if they don't, they will be forever correcting and returning travel expense forms, timesheets, and authorizations. They also know that it's no sin for you to admit ignorance about paperwork, and they will be flattered that you value their time enough to ask how, first.

- *Meetings.* As soon as you start your job, you will be asked when and where the meetings are going to be held. Some work requires daily review meetings, some can be done with weekly ones. You usually have some latitude in setting up your own system. Just remember that, although meetings have valid purposes (such as allowing you to benefit from group synergy in problem-solving), they are also extremely expensive and dangerous interruptions to the workday. Figure out what one hour of meeting time costs and keep the number permanently on the blackboard. Never hold the start of a meeting for anyone. Always start with specific objectives on the board. Always end with a list of information needed for the next meeting. Don't hold anyone in a meeting that doesn't concern them. I'll return to the subject of meetings later on. Meanwhile, take a look at any of the books on effective meetings.

- *Dress.* Yes, you may have to change the way you look. This is a non-issue for technical workers. The normal engineer, scientist, and first-level manager of same is usually not concerned with clothes or appearance. In fact, some of us have been known to gravitate toward companies with no dress code, tolerance of odd lifestyles, and disregard of the ancient rules about having food, pets, radios, or children at workstations. However, if you're a manager, you run the risk of increased contact with customers, sudden lunch meetings, or even having to speak to outsiders. We all have an understanding that it's quite reasonable to avoid "businesslike" dress, but this understanding does not extend beyond the front gate. If your new position involves more travel, customer contact, conferences, or trade shows, you need at least one middle-class business-person uniform. Wouldn't it be a shame if the only reason you didn't get to go to that very important evening meeting with the capital supply boys was because your boss was embarrassed by your rags! This whole area creates controversy, because there are significant regional and national differences in what's usual, and because women are expected to dress in conflicting ways, i.e. look "nice" while working in a dirty or hazardous environment.

- ◆ *No promises, no deals.* All at once you're in a position to deal for resources with other groups, with peers, and with service organizations. You will see all kinds of opportunities for getting your wish list done by promising things to your staff in return for heroic efforts. Try to remember that you really won't know what you *can* deliver until you've been at it for a while, so keep dealing to a low level. Also remember that burn-out is directly connected to empty promises and can hurt your staff.

- *Focus.* So many things are happening that it's easy to forget to concentrate on the primary project objective. Anything you do that is not connected with getting that specific job done uses stolen time. As you will see later, planning means working backwards. Start with the objective, construct a plan for getting there, find the resources and tools to fit the plan, and make it possible for the work to get done. Empire-building, attacking other managers, and building external reputation all do not contribute to the project. Also keep thinking of leverage; at any given time, how are you causing more work to be done if you weren't there?

## SUMMARY

The first few weeks of a new managerial job require that you define objectives, gather information needed for a plan, meet your staff, ensure continuity of work, find your place as the head of your group, and establish your relationship with your boss and within the company organization. Expect to lose sleep, weight, and generally be stressed. Test the controls and show that you are, indeed, a manager.

---

**What Do You Know After a Few Weeks on the Job?**

- Who
- How
- When
- How much
- That you can or can't
    - Delegate
    - Communicate
    - Direct
- What's in the Manager's Toolbox

---

# Chapter 2

# The High-tech Difference

*High tech is the production of innovative things.*

The high-tech business invents, develops, produces, and sells innovations. A successful company is structured, top to bottom, to manage innovation effectively.

A business that innovates is different from other businesses — it hires different people, responds to different market forces, and does a number of things that would make no sense in noninnovative businesses. Your job as a manager is built uniquely around some requirements of your high-tech world. This chapter explains what these requirements are and how they affect what you need to do.

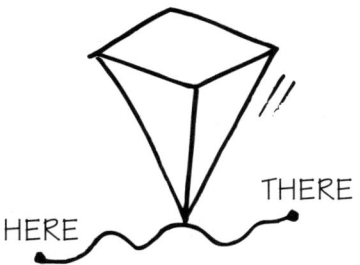

HIGH-TECH NONEQUILIBRIUM

Your high tech organization is:

- Dependent on continuous innovation
- Extremely responsive to market changes
- Sensitive to the smallest technical decisions
- Capable of rapid reorganization
- An employer of highly skilled specialists
- Knowledge-hungry
- Lean in structure, fat in technical smarts

### High Tech in Low Tech: A Division of a Big Company

If you're in a high-tech division of a much larger outfit, you may find that this description of a responsive, loose, lean organization doesn't seem to apply. You've got to deal with all the *corporate inertia* that fights any kind of change. It's a serious problem and always has been. Many good people have concluded that only small companies and start-ups can compete effectively in high tech. This is not true. A big, clumsy, hogwash-dripping company *can* compete, if not efficiently, by virtue of throwing its massive resources at a problem. It may spend ten times as much to do a given job as a start-up, but it can get the job done. It is, however, a very different kind of environment, and you have to adapt to it. The megacompany sure won't adapt to you! Chapter 3 gives some pointers.

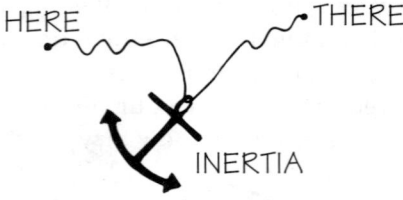

Let's look at some of these characteristics and what they mean to you on a daily basis.

## CONTINUOUS INNOVATION

Any product passes through the stages of newness, maturity, and obsolescence. In our world, there isn't much of that second stage. This is a high-tech difference. Electronic devices, for example, evolve so quickly that last year's part, despite wide availability and debugged production, offers fewer features for the dollar than this year's part.

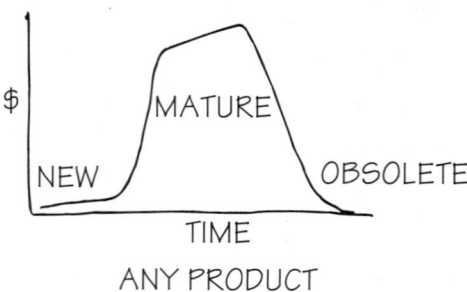

♦ When competition is fierce and technology advances rapidly, products can be obsolete long *before* they are available for sale, which is a disaster. Everything your company has spent on development will be wasted if your competitor hits the street before you do. Inventions and breakthroughs anywhere on the planet immediately help or destroy your product. So the pressure is always intense to be out in front, on the leading edge, coming up with really new stuff. This is product revolution instead of evolution, and it's a tough game.

---

**Crushing the Innovation**

Right now, high tech has a new problem brought about by the combination of a slow economy and short-sighted investors. The problem is that innovation looks like a bad investment, because the payback period from mature product can be too short to justify the enormous development cost. The resulting slowdown in R&D schedules doesn't only slow down product development, it destroys whole industries. The product is obsolete before production.

Meanwhile, the truly revolutionary ideas can't get sufficiently financed because there is no way to project the business potential of a breakthrough. Flat-panel displays are a good example of guaranteed business potential failing to open purses. Everyone is convinced that flat televisions, computer screens, even highway signs, are going to displace vacuum tubes. In the United States, where all of the most promising technologies in this area were invented, the level of investment in display development is tiny and falling.

---

All this innovation can come from inside or outside your company. Although it's true that your company doesn't have to do all the creative work, competition usually means that there's no time to copy another company's work unless you have a manufacturing advantage or some other capability that the innovator doesn't have.

Texas Instruments, while being a major innovator in hundreds of products, also had a reputation for being able to get in late and outproduce the companies who originated mass-market products. This strategy is one we often associate with Japanese industry, but it has been used successfully in this country. It works well as long as the market is big enough and the product has enough longevity. If there is enough room for competitors, however, other (smaller, quicker-responding, more efficient) companies will jump in and erase everyone else's profitability. This happened to TI in calculators and digital watches. In fact, there is always a background of "virtual competitors" who spring out of the ground to undercut you as soon as you try to recover plant or development costs.

In semiconductors and electronic devices, a subindustry of product copying (theft?) has been created. Sometimes called *retroengineering*, this process involves specialists who use extraordinarily sophisticated analytical tools to peel apart and copy products as complex as microprocessors or as simple as glues and coatings.

Because patents and other protective devices don't work well, especially in electronics, innovation becomes more or less public and fair game for knock-offs. This makes it even harder to justify spending money on research, because it's risky. Copying a working product carries far less risk.

When the innovation comes from your own shop, it's usually unpredictable, unplannable, and can't be forced. The manager's job is to make it possible for innovation to flower, to recognize it when it happens, and to reduce it quickly to practice — that is, make a product from it.

In my own experience, I've seen three Nobel Prize ideas get buried and forgotten in lab notebooks because managers couldn't understand their importance or were unable to promote and develop them. Although the Nobel Prize is not a great indicator of commercially viable ideas, I think that you can get the idea: *An underdeveloped, secret, or back-burnered idea is pure waste.*

```
                                 ...·THERE
           HERE      *....·´  ·
             ·...·´¯¯¯¯\.
                        \.
                         \.SOMEWHERE
                            ELSE
         *INNOVATION,
          CHANGE
```

It's asking a lot to expect us both to recognize good ideas and have enough perspective and skill to develop them. We may know almost nothing about business, markets, and promotion. Generally, scientists and engineers embrace any idea that seems to be new, innovative, or clever. ◆ Where we fall down is in the ability to follow through on the development process.

How to handle innovation:

- Keep an open mind. Don't make snap judgments about new ideas, like "That's too simple to be really new." Let everyone in on the brainstorming, and keep personal prejudices under control.

- Always encourage the idea generators, even if they sometimes miss the mark. Creative people want to be respected for their creative *ability*, not necessarily their concrete achievements. They become secretive, or worse, when they think you're not appreciative. Secretiveness means you do not hear about potentially great work. "Worse" means that your best people may depart for greener pastures, with their ideas. Behind every successful spin-off is an unappreciative manager.

- Treat ideas that seem irrelevant, or currently useless, as if they're paintings that may someday become valuable. Make a point of requiring good notebooks and record-keeping as a routine practice.

- When you spot a winning concept (provided you don't run with it yourself!) know how to promote it in the organization. Remember that your boss, and his boss, are further away from understanding the technical detail, and promotion is necessary, even for the best ideas.

## RESPONDING TO MARKET CHANGES

Everybody has met the "inventor" whose stock in trade is the secret, super-duper revolutionary gizmo. His idea is so wonderful that it can't be disclosed or studied . . . and of course, it's nearly always trash. (On a number of occasions I've sat in with patent attorneys to evaluate the merits of what turned out to be perpetual-motion machines, sources of energy from nothing, and other high secrets. I even have a letter asking for specs on parts for a proposed time machine. All of these inventors started with a valid, if obvious, market concept. Unfortunately, they had failings in other areas.)

High tech creates its own market some of the time, but it also has to mesh with other developments and real needs. If it doesn't, nobody will want it. Secret hatching of new things is a fine idea, but in practice, lots of people have to know what you intend, why you think anyone wants or needs it, and how the market would accept such a thing. The market has to be ripe for the product.

On the other hand, many of us waste time and creativity trying to respond to commonly accepted market demands that turn out to be false. It may be a premise that seems reasonable (at least to engineers), such as "Every household will need to have a personal computer by 1983 to keep recipes, do schoolwork, keep the checkbook, and . . ." (a common premise around 1976), or "Everyone will buy an energy-efficient light bulb" (1978), or "Systems makers will all need chip fabrication capability" (1979).

The high-tech relationship with the market is based on the following elements:

- Close customer interaction
- Close observation of competitive technology
- Very little test-marketing
- Lots of common sense and perspective

How to know your market:

- Talk to people *who are not in your field*. Your coworkers, the trade press, the marketing department — all may be infected with some attractive delusion. The end-user of your product *isn't* in your business. Trust the user.

- Talk to your actual customers, even if it risks the ire of the sales or marketing folks (see the box below). You might be surprised to know that customers have detailed wish lists that can get garbled by passing through too many departments on the way to your desk. The payoff can be tremendous.

- Brainstorm with your own group. They may expect to be told what to work on, but they are also privy to informal, direct, information channels, such as friends in other companies. The grapevine, as erratic as it is, transmits information you can't get any other way.

- Have your own opinion. Use your common sense. Some products are obviously inevitable, others are questionable. Half the history of the personal computer is that of a herd mentality trying to generate undifferentiated products in a saturated market. If your company is jumping in with a "me-too" product, there may not be a market left by the time you get it shipped.

---

**On Your Toes . . . and Off Theirs**

The marketing department does work that overlaps yours, and in an ideal setting, marketing staff should be working closely with engineering, R&D, manufacturing, and service. In reality, many of us don't even have a good idea of what marketing's job is, and many marketing people believe that they don't need anyone's help. Thus, whenever you (in innocent good faith) go to a customer to find out what they need, you stand a good chance of stepping on the toes of marketing, sales, or both.

Sometimes, to control information flow between your company and its customers, there may be formal policies that prohibit contact by anyone except specified personnel. If this isn't stupid enough, add in a little empire-building and ego problem in the marketing area, and you get the complete experience of an organization poised to make some totally unwanted product.

♦   You should try to get friendly with any department that is chartered to contact the customer. Offer your advice, facts, and figures, relevant bits from trade magazines, and anything else that could make their job simpler. I've even arranged for little tutorials between the development staff and the marketing staff, for mutual benefit. You need to show that your technical abilities are vital for the interpretation of customer needs and that you won't do anything scary when you do talk to the customer.

## SENSITIVITY TO SMALL TECHNICAL DECISIONS

A low-tech business succeeds when upper-level business or strategic decisions are correct. A high-tech business can be made or unmade by relatively trivial technical decisions made by a draftsman, a purchasing agent, or a junior engineer. I think that you can measure the "high-techness" of a company by the extent that this is true. This awful sensitivity comes about because:

- Product development is too fast for second chances
- Specialization means not understanding the product
- Fatal errors can't be spotted early

So, if an engineer specs a component without knowing that, for example, there's going to be a large induced RF current in a ground plane caused by another board, the board will work beautifully by itself, but the product will crater. A wrong choice of materials for an innocent little connector can contaminate some other part of an analytical instrument. And so forth. A common problem these days is having only one pass at making a critical semiconductor chip without also having the ability to simulate or breadboard its function.

How to control small, important decisions:

- God *is* in the details. Change your attitude about what's trivial and not worth your attention.
- Make sure *everybody* knows exactly how their little bit of work will be applied in the end product and how it will be used by the customer.
- Create an environment in which *everybody* feels that the work they do is critical and therefore must be flawless. Make sure that nobody believes that errors are either invisible or will be caught by mysterious others, such as you. On a modern production line, for example, *anyone* can hit the "stop" button.

## CAPABILITY FOR RAPID REORGANIZATION

Most of the time, you can't even *detect* the organization of a really hot little company. Everybody appears to do their jobs without detailed direction, and something like an organizational chart is invisible, irrelevant. This is because skilled, professional people are self-motivated and understand their jobs better than anyone else, including their bosses. An invention or a breakthrough can cause the whole company to shift gears, and change focus. Similarly, an order from the top to drop a line of research and go after something else can be implemented without memos, objections from vested interests, a waste of expensive tooling, and all the other impediments clinging to low tech.

How to be fast on your organizational feet:

- Learn to express instant love for new directions, no matter how weird or vague. If you hang back or grumble, you slow down your group's response. If it's a really bad idea, complain to your boss, *not* your group.
- Don't be stuffy about formalisms like mission statements, new organizational charts, new account numbers, and proper facilities. Go with what you've got, and give the slow-moving overhead critters lots of time to ponderously do their jobs. I know people who have never had a business card that described their current jobs, and who don't worry about it either. If, however, your boss or the organization itself is stuffy, you have to live with it.

## AN EMPLOYER OF HIGHLY SKILLED SPECIALISTS

You may have, or be, the one person in this country who is a leading worker in a particular and vital area. The *front-end* part of a high-tech company is composed of irreplaceable specialists. Some companies are entirely dependent on a single key player. Even in less-extreme cases, unusual, arcane, and rare skills are necessary to success. Staffing up, replacing and keeping staff, and just understanding job skills are much more critical and difficult in your company.

How to live with exotic and odd-ball staff:

- Every time you want to bite some geek's head off (an interesting turnabout), think long and hard about how long it could take to find a replacement. Threats and terminations are a poor excuse for management skill.
- Don't be embarrassed to expose your own ignorance in pursuit of finding out what your kingpin actually knows or does. If the expert walks out tomorrow, you had better be able to survive without those rare skills.
- Extreme job, extreme person. You have to tolerate, respect, and try to understand some unusual coworkers. If you like homogeneity, you're in the wrong game, Bub. People who appear to be okay when you are peers have a distressing habit of becoming intolerable when you become a supervisor. Watch out for this effect.

## KNOWLEDGE HUNGER

New products come from basic research, applied research, development, market forces, and inspiration. Your outfit is constantly involved in reading, conference-going, consortia, and moving and sharing information in dozens of forms. The difference is that knowledge gathering is done all through the organization, right down to the bottom layer, so that everyone has some responsibility for it. Because small decisions can have giant effects, missing a minor but key development can turn into a big problem. The following methods of staying current can help:

- Make up a list of where the knowledge you need is likely to come from. Portion out the list to your group, giving each member a specific area to cover. Don't, however, give all the attractive stuff — such as overseas conferences — to one person.

- Don't waste time turning this list into memos and reports. (Your boss, however would like some punchy abstracts.) Cut a little meeting time for "the news" every week, and encourage on-the-spot discussion of significance.

- If you want to increase the utility of information gathering, stuff a key-word-searchable database with your gleanings, preferably one that is company-wide and connected to a bulletin board or internal news service. Remember that not everybody can leave their work areas to go to conferences, or even to the library, but that nearly everyone can get to a terminal.

## LEAN IN STRUCTURE, FAT IN TECHNICAL SMARTS

When the bench workers have advanced degrees, are self-motivated and skilled, and can measure and correct their own work, what's left for a manager to do?

In fact, the traditional directing and controlling role of the middle or lower manager is vanishing in better technical companies. Even the specialized planning and budgeting job can be done to a great extent by easy-to-use software. The desk-bound, paper-shuffling overhead-eating manager is extinct. And not missed!

The structure of a high-tech firm looks more and more like a democracy, with important decisions being made by individuals and accepted by upper levels. Everyone can generate reports, analyses, and projections. This is not to say that two groups, such as marketing and engineering, won't still fight over the design turf, or that upper management won't blow off sound technical judgment as always. ♦ What's new is the *extent* of the democracy, and how important it is.

The manager's job runs more toward facilitation than anything else; that is, making it possible for staff to get work done with minimal interference and maximal resources. Sometimes it looks like management by magic.

How to keep from being an extinct manager:

- Concentrate on the job of helping other people to do their jobs. Become the expert on all those dull, nitty-gritty organizational functions, but don't forget their relative unimportance.
- Protect your staff from excessive, time-wasting paper.
- Use your authority to bring more resources into your group. But don't get greedy.
- If it's a democracy, you're the coordinator.

## HIGH-TECH MANAGEMENT CHARACTERISTICS:

High-tech management is:
- Flexible
- Loose-reined
- Not shy about ignorance
- Willing to have smarter subordinates
- Able to communicate with distant planets
- Networking
- Willing to fail every so often
- Sensitive to burn-out
- Interested in fame and fortune

Say what?

This picture is all wrong from the point of view of traditional management! The all-knowing, all-controlling, never-failing Organization Man can't make it in our game. The concept that it's okay to manage tasks you can't understand is bad enough; adding the concept of failure as a fact of life would get quite a few factory managers annoyed.

Quite a few high-tech business practices are nontraditional. Spin-offs are usually positive developments rather than mutiny by disloyal staff. Opening a company's most vital plans to consultants and other outsiders is also considered to be proper. For better or worse, large, successful high-tech businesses have appeared and prospered without seeming to need the baggage of more established companies. Let's take a closer look.

### Be Flexible

The job always changes, the product changes, the technology changes. You change with it.

### Keep a Loose Rein

The creative ones don't punch a clock. You don't order brilliance.

### Admit Your Ignorance

The managed know more about the work than the managers. 'Fess up.

### Be Humble

Some of your staff members are even more famous than you. So why are you the boss?

### Communicate

Closed doors and secrets mean bad products. Pass it on.

### Network and Participate

Technology is global. Knowledge is not all in one location. Even if Dudtek is the "industry leader," it won't stay that way without lots of help.

### Be Failure-tolerant

That's research. Sometimes you have to decide when to quit. Sometimes it really hurts.

### Help Prevent Burn-out

♦♦   Talent is not replaceable; it must be protected from damage.

### Seek Fame and Fortune

Again, the company isn't the whole universe. Unless you've got some dreams that include money, your own company, and similar things, you may be too meek for this business.

---

## Us and Them

For several years, the trendy thing has been to copy offshore — especially Japanese — practices, which, when reduced to jargon and buzz-words, give us an image of slavish, egoless operations run by committee. If we see obvious success in every high-tech manufacturing application, we should feel that their management methods are worth study. However, I believe we're looking for the easy version — the one that doesn't involve going there or learning the language. As a result, we get, by and large, a comic book description that leaves out all the important stuff. Then we get frustrated because it doesn't look as though Americans could follow the recipe, and we mutter things about how "they can't be creative."

Hogwash. High-tech success is a matter of getting many, many ducks in a row, from R&D through manufacturing, raising capital, and sales. The Japanese have an unusually appropriate environment that they built by paying close attention to the obvious failings in other systems. At the front end, where the creative work happens, most of this chapter applies equally to "us" and "them."

## THE HIGH-TECH MANAGER

The high-tech manager is:

- Technically competent
- A solid communicator
- Highly motivated
- Resilient
- Mainly, a good manager

Other than the concept of technical competence, these requirements are cause for some concern, at least for many scientists and engineers. They often represent clear areas for personal development. To be a manager of engineers or scientists, you have to be able to understand the important aspects of work whose details you cannot follow. In addition, you have to be able to translate these ideas into English and communicate them to other managers in a way that is relevant to *their* priorities.

You have to be willing to go on gut feel and make nonquantitative assessments of work in progress. You have to understand the end use of your group's work and its importance to some user or customer. You have to *relate* your work to other parts of the company. If you're in research, you have to understand the methods and capabilities of manufacturing. The manufacturing manager must know a great deal about sales, service, and the customer.

You have to be motivated in order to motivate others, or even to figure out what their motivations are. Many of your people would do their jobs almost without pay, for the chance to work with the tools or with the other people. Sometimes these strong motivations are counterproductive, and it's your job to keep individual and company goals roughly aligned.

### Who Gets to Be a High-Tech Manager?

Nearly anyone can become a high-tech manager. My first boss in an engineering job was an ex-minister. The project was an observatory satellite, and there were about 20 engineers working for Rev. Boss. He had not a clue about the work being done, but it didn't seem to matter. Since then I've seen engineers, liberal arts types, undegreed technicians, MBAs, and even a handful of Henry Ford-like self-educated managers. The great corporate pinball (or better, Pachinko) game bounces, bumps, and spins all kinds of people into lower-level management. Sudden awards of large contracts, mergers, up- and down-sizing, transfers, firings, and the continual turmoil of industry all ensure that managers, above all others, need not have rigidly defined backgrounds. You may find that galling, but the evidence is plentiful that a manager can be a winner despite "wrong" qualifications — or a complete dud despite the best ones.

You, on the other hand, are going to be a *good* high-tech manager. The fact that some of your fellow managers, maybe your own boss, are non-contributors or even deadwood shouldn't get you down. If anything, it should represent an opportunity for you to stand out by being better at this job than they are.

### Technical Competence

Technical competence isn't as easy to define as it once was. The rate of change in most technical areas is so rapid that your skills tend to become obsolete unless you're totally immersed in leading-edge work. For nearly all of us, being good at management means giving up some technical currency. It may mean getting your information from abstracts and summaries, rather than from the full text. It also means having your staff know more than you do about the details.

You can't afford to lose some skills, however. You can never give up being quantitative and analytical. You have to know how to attach numbers to concepts and how to attach concepts to each other. These days, a firm grasp of matters statistical is mandatory. Hang on to your problem-solving hat. Exactly the same analytical approach you needed in school is needed on a daily basis in management.

You also have to keep current in information-handling skills. This could mean versatility with data networks, or it could mean the ability to access technical and business literature. You need to know what's going on, who's doing what, and what it means. The only way to do this is to be a regular user of information resources.

Fundamentals are just that: the foundation of everything else. Have you forgotten everything you knew about the love life of the complex variable? Has all that stuff on control theory washed away? Don't know an eigenfunction from an eiderdown anymore? Don't worry. As long as you remain on close personal terms with freshman physics and chemistry, you can recover the rest when you need to. You'd be better off to let the seldom-used 80 percent of your vocabulary rust away than to forget how to read!

Sometimes technical knowledge *is* critical to the job, though, and you have to continue to be a technical contributor while adding management responsibilities. In those cases, you'll have to become selective and more efficient — or there just won't be enough time in the day. What this means is that you'll read the abstract instead of the article; you'll rely on other people's summaries and analyses; and you'll tend toward designed experiments instead of brute force. It's the concept of *leverage* again. Your time costs more, so it has to be more productive than it was.

### Communications

Think of communications this way: If you can't communicate really well, you're stuck with doing everything yourself. You're also a poor manager.

```
ME •——←——[ MEMO
            MEETING
            LAN
            LUNCH  ]——→—— • THE
```

COMMUNICATIONS

Communication is difficult, especially for technical types. It implies all sorts of indeterminate and soft concepts like personalities and psychology and motivation and dealing with people you don't like or are afraid of. I'm sure that you know more than a few intelligent, otherwise competent people who go into shock every time there's a routine group meeting, or who struggle for days to write the simplest memo.

What has happened, unfortunately, is that we've learned to substitute paper and electronic mail for human-to-human contact. Although the electronic office has the potential for being an interactive and lively place, it also generates noise. The number of bits goes up, but the information content continuously decreases. In large outfits, the bulk of everyone's time is sometimes spent on generating or redirecting unread, unnecessary junk mail. When it gets to that point, the organization has become nearly useless.

Everyone would agree that the paper blizzard originated in Washington and spread through large contractors first. Electronic excess, on the other hand, arose at computer makers like IBM and DEC before spreading everywhere else. One of Stein's Laws (#32), says that:

*The volume of superfluous communication is inversely proportional to the health of an organization.*

Communication is getting your ideas understood by others, and understanding their ideas. It's obtaining agreement on the best course of action. It's ensuring that no work is duplicated and that results are known as early as possible. It's keeping all the plans and schedules meshed. In fact, it's the force field that holds your group, and your company, together.

To do your job, you will have to communicate easily and frequently. If there's something the president should know, you have to be able to find the right way to get the message delivered. If your worst worker needs to be encouraged, you need to know how to do it. None of this is easy, and you probably won't have the luxury of much formal study!

### Supervisory Skills

I've said (several times) that supervision in high tech is different for several reasons. You've got highly skilled people, odd personalities, pressure from everywhere, and you're always doing something that has never been done before. The good part is that supervision of intelligent, motivated people who enjoy their jobs *can* be pleasant, bordering on wonderful. If you do your job of planning and facilitation, they'll do theirs, and everyone wins.

The base for your skill is thorough knowledge of your organization and, the business that it's in, as well as a feel for how people interact on the job. From this base, plus a little reading and observation, you develop strategies and techniques that appear to work for you. You don't copy somebody else's style, and you don't copy another company. Your situation is unique and requires unique tools.

One general rule, though. Those who work for you need to see you at work. Your people need to see managerial and supervisory skills in action. They need to see planning, decision-making, resource allocation, and advocacy. If they don't see you doing these things, they may get the idea that you don't know how or, what's worse, don't care.

## THE MANAGEMENT ZOO

There are many different styles of management, none of them particularly right or wrong. Your own personality mixes with the company personality (the *culture*) and the individual personalities in your group. The result either works or needs changing.

I've seen gross, venal bosses function successfully among delicate scientists. I've seen autocrats, abusive, insensitive tyrants, and people with severe personal hygiene problems run smooth and productive outfits. I've also seen compassionate, intelligent, and open managers work well. I've seen all different types fail, too. What follows is a little bestiary of management styles. See if you recognize any of these animals.

### The Guru

This beast's dominion is based on knowing more about the technology than anyone else. Presumably, all the important new ideas are handed down from above. The group's function is to carry out the guru's wishes and thereby enhance the guru's reputation. It is fairly common for our sorts of businesses to hire gurus deliberately, for their marketing utility and their presumed ability to transfer technology quickly. Guru-induced problems are many and obvious. The guru frequently cannot relate to peers in management and can only supervise submissive or undeveloped staff. The group's creativity and initiative is obliterated and tremendous frictions ensue. In a transient situation, such as graduate education or a quick project, the arrangement can be made to work. In business, however, it's generally bad.

### The Master

This is someone who derives all of his authority from his placement in the organization. This is the classical manufacturing boss: the autocrat. Because discussion and communication is so vital to developmental work, an autocrat is, to some degree, hindered by the elevation he has acquired. He has to balance the remoteness of authority against the necessity to be a participant. When the group consists of intelligent, educated people, most of whom have had several jobs and many bosses, the master-slave relationship doesn't cut it, because the basis for respect is not there.

### The Provider

This animal is skilled in the administrative arts and is recognized for it by his peers and those higher up. Because he has the ability to handle funding, support services, facilities, and planning, he generates respect from the group even when his technical qualifications are thin. A good provider makes everybody's job easier, and he has a concrete and defensible job. He is not considered to be a threat by creative people, who generally are not desirous of taking on administrative tasks. The disadvantages are related to technological weakness. The provider may not understand results well enough to present them to higher management, and he cannot be relied on to serve as the focus for essential technical networking.

## The Motivator

Because the failings of autocratic supervision are most obvious in technical work, more emphasis is given to training in motivation, persuasion, and "people skills." The value of motivation is not a new concept. These days we're more likely to say that a manager is useful only when he's above unity gain—that is, when his presence causes more work to be done than he could do alone. The best way to have "gain" is to be a good motivator so that your crew assumes more initiative and requires less of your time in minor decision-making.

Motivation and selling, however, are similar activities. The danger of concentrating on motivation is that you can, and must, sell ideas to yourself before you can credibly sell to others. Often this will result in some loss of critical perspective, causing you to lead your group too far down blind alleys because you have convinced everyone that a concept is too good. Just as bad, you could give your manager an unreasonable picture of what you propose to deliver.

Regardless of whether it's possible to turn an introverted, nerdy engineer into a personable leader, many thousands of people earn their living producing courses, tapes, and books that are used in corporate America for just this purpose. Unfortunately, skill in this area can't be acquired easily and probably involves wholesale personality changes. Some people can be motivators, and some can't. If you happen to be one of the latter, you should reserve your best efforts the way you would the supercharger on an engine — for limited use in specific situations requiring unusual results. Otherwise, you can eventually burn your staff out, lose their respect, and become known for over-promotion.

## One of the Boys

A safe and tempting style of management, especially for those who are newly promoted, is to act as if there were no distance between the group and the manager. The latter is still "one of the boys," and decisions should be made in a democratic, collective way. Researchers tend to like this arrangement because each member of a group is likely to be a contributor, have opinions, and be willing to be reasonably open about what is and is not known. Researchers also feel comfortable when the boss is active technically, and the boss usually has some fear of being excluded from the laboratory. Democracy is also attractive to those who are a little timid about making decisions and like the anonymity of collective ones.

♦ As a result of Japanese example, collective decision-making has become a hot idea and has even been applied in manufacturing in this country. In the best cases, it eliminates one or more layers of management, increases productivity, and improves the coupling between the product, its engineering, and business decisions.

- Collective decision-making is counterproductive when responsibility is really focused on the manager, when unpopular changes in direction need to be taken, or when it eliminates managerial skills from the job description. The working week is short enough as it is, with meetings and other zero-worth activities, and a manager who is trying to do a decent technical job *and* play manager, is forced to shortchange one or the other. It is also a serious question whether our culture of individualism even allows us to work this way.

### Angels and Devils

Angels and devils are those who, for some reason, radiate love or fear and are loved or feared by their employees. I think more of these people are found in fields remote from technology, but there are enough of them around to warrant description.

It takes more than a weekend management seminar to turn a mechanical engineer into Glinda or Attila, but we should recognize that few of us are charismatic and natural leaders. Some of us bring a cold, damp wind into the room. People will respond to your personality by establishing a clear relationship with you, which may even be highly productive.

### And the Winner Is . . .

In technological organizations newly promoted managers tend toward one of two extremes: they become either *autocrats* or *one of the boys*. Being an autocrat is tempting if you've always wanted to exercise authority. Being one of the boys is a safe route if you're unsure of the value of your decisions. The research lab, in particular, is prone to democracy and collective decision-making. Then, too, we've got the Japanese example in which group consensus seems to work well in high tech.

You must pay attention to other managers' personal styles, so that you avoid trouble, but you don't need to emulate anyone. Chances are good that you can't do it, anyway. Being scrupulously nice doesn't mean that you get Louise's angelic reputation; you might just be written off as weak or uncritical. Conversely, aping Mike's tough and decisive manner won't necessarily get you more respect, but it will get you avoided.

Your own style will evolve more or less naturally. If you want to explore styles in a more formal way, see the reading list.

## MOTIVATION

Read Dale Carnegie. Seriously, folks. You may not want to be seen with it, but *How to Win Friends and Influence People* is still the clearest description of what makes most of us put in that extra effort. On the other hand, the sort of material you'll find in a management curriculum these days runs to theory you'll gag on — unless, of course, you're interested in becoming a management theorist!

After you read about what motivates the normal worker, consider the rest of us. The software-writing, candy-eating night creature glued to the screen here and at home. The hollow-eyed chemist trying to get a dozen simultaneous papers into print, all on the same data. The hermit down the hall who is convinced that unless he can further reduce a particularly complex equation, unknown parties might reduce him to a point and send him off to infinity. (This last example, sadly, is real. He was a chip designer in Romania, where many people had good reason to become severely paranoid. Infinity wasn't the destination, but he *was* sent away.)

Consider *your* motivation. Why do you do your job? The money isn't spectacular. Do you actually like having no job security, no permanent home? Do you like messing with as-yet-unknown hazards on a daily basis? Do you get a charge out of seeing projects screech to a halt just when they were starting to look good? What makes you do it? See if you can write a list of your own motivators. Discuss it with someone you trust and add their opinions.

This could be a depressing inventory. However, you're not going to be a good motivator of other people unless you have some self-knowledge and are excited by your own job.

Now, Mr. or Ms. Manager, you've got two jobs to do. The first is to motivate yourself to go nearly all-out on the present job, and the second is to get everyone else into a similar state. The reason I say "nearly all-out" is that an excess of zeal can be destructive and can contribute to burn-out: yours and theirs.

## RESILIENCE

You have to be able to ride through all sorts of grief. If you become too attached to a concept or a project, you may follow it too far, to your regret. You can also damage your group and your organization by inflexibly following an outdated plan. Spend some time with an experienced manager. Ask how many times they've had a cancelled project, lost an employer, or changed career. You'll find that people are either bitter about frustration on the job, or they have learned to adapt. You will be a much better manager if you feel comfortable with change and all those mighty forces beyond your control.

## YOUR JOB

1. Why is your company high tech? List those characteristics that differentiate it from a company you regard as low tech.
2. If you're in a startup company, you probably have only one product under development. Why do you believe that your company has an advantage over

larger, established firms? Is it technology, marketing, particular "superstars," or something else?

3. How long will your product persist in the marketplace? How often will you personally be involved with a major product change?

4. Have you ever been a user of your company's products? If not, how often do you talk to customers?

5. If your product is dropped, would you consider going to another company so that you could stay with the product? Would you go out on your own?

6. Rank the characteristics of a high-tech manager in your own order of importance.

7. Give yourself scores of A,B,C, or F on each characteristic.

8. In your new group, how many members can make technical errors without either slowing development or ruining the project? Is there any particular member who must be right all the time? Is it you?

9. As a boss, are you an autocrat, a democrat, or something else?

10. Estimate the fraction of your time that is spent on communication. Include meetings, preparation for meetings, electronic modes, receiving status reports, walking around, and lunch.

### Scoring

If you can answer most of these questions, you have a good idea about the high-tech swamp, your boat, and its captain. If you have trouble finding answers, you may have trouble navigating, because you don't know where you are or where you're headed.

Questions 1 and 2 concern what your company does and where it fits. The answers are your perceptions and are open to debate. Try discussing your point of view with someone else.

Question 3 asks you for historical perspective. ♦ Knowing something about product cycles and longevity will allow you to put together more meaningful plans and to understand the difference between *fast* and *slow*.

Question 4 is intentionally rude. You need to know the end use pretty thoroughly in order to make a good product. If you have never used the product, you need constant contact with those who have.

Question 5 measures your degree of involvement with the product or the technology. A good answer is "maybe." If you feel less committed to the product, you may not be sufficiently motivated for the job. If you're ready to take it outside and run with it right now, it may be a sign of impending loss of balance, or it may be time to go for it!

Questions 6 and 7 should restate some of this chapter's material, but in your order of importance. Each of us will come up with a different order, because our jobs and styles are different. What's important, however, is whether your high scores match the characteristics you consider most important. If not, you've just defined skills you need to develop. Also, if you rank something at the bottom and also say that you're terrible at it, you're identifying a probable blind spot. Work on that, too.

Question 8 asks you how critical you believe the work is. If you think that nobody can make errors or that *you* have to be right all the time, you're headed for problems. The problems won't necessarily be with the work. They will be with your group.

Question 9 is another one you should discuss with other people — your group, your family, your friends. You may find that not everyone sees you in quite the same light.

Question 10 gives you a benchmark to contemplate later on, when you try to find more hours in the day. The answer should appear excessive now and skimpy once you have a better understanding of the management job.

# CHAPTER 3

# Tiger Team, Skunkworks, or What?

### CULTURE

So here you are, embedded in what appears to be a company that does high-tech business. The difference between being "embedded" as a fly is in amber, and being "embedded" as most valued employee depends on how well you understand the organization around you. If you want to get things done, you have to know how the outfit works, who to go to with problems, what is *really* expected from you, and how to know whether you're doing your job in the company-approved way.

Your company has a unique way of doing business. This "personality" is usually called the *company culture*, when someone bothers to put a name to it, and the variability of it is astounding. If you've only worked in one organization, you may not have noticed this unique way of doing business, and you may not think about it very much. However, as a manager, you have to understand the fine structure of your culture better than you know your own family, and you need to develop new skills if you want to get the most from your resources. Otherwise, you'll be as immobile as that fly.

Old, large, and well-established companies have cultures that are easy to understand and that don't change quickly with time. Often you can get a good feel for them in a short time, and sometimes you can study training materials or read the personnel handbook for hints. Very large companies get the urge to codify their idealized cultures in some sort of Holy Writ. The concepts of sticking to "channels" — that is, paying attention to the formal organizational chart, not making waves, and working neither too fast nor too slow — sometimes apply. Knocking yourself out with a 16-hour schedule isn't usually required. Unless, of course, your old, established company is seeing some competition or has gotten trapped on the leading edge.

Most high tech is a lot different: Speed is critical, and a channel is where you find it. In fact, where you work it's possible that nobody has ever seen an organizational chart! (Every place I've worked has deliberately suppressed all the detailed charts and rosters because of the threat of headhunters and corporate espionage.) High-tech companies range in their operations from near anarchy (sometimes seen in basic research labs, "Skunkworks," and especially software development) to the uniform-and-haircut level (seen in manufacturing, the IBM of yore, and one-man shows like EDS). The culture may be unwritten and unspoken, may vary from division to division, and change from good years to bad.

♦ All of us differ in our ability to understand the company culture. Technical personnel tend to ignore even the most obvious and rigid cultures because we don't perceive any relevance to our jobs; after all, we're not in the game for promotions, and we don't consider ourselves as managers, who would have more reason to worry about this kind of thing. Besides, defining our personality and lifestyle *isn't our employer's business*. If only that were true.

♦ The culture puts limits on what you can do, how you do it, and how fast you should work. It defines the official organizational structure, the invisible organizational chart, use and mobility of resources, rules for contact with the rest of the world, ethics, personal behavior on and off the job, and even political and religious preferences.

---

### Big Brother Is Employing You

Some companies really make a habit of imposing their culture on one's personal life. Here are some of the milder policies that have this effect:

- No promotions for unmarried employees
- Demerits for those who take their earned vacation
- Dress codes

They get more interesting, and often illegal:

- Make it difficult for the handicapped
- Keep women out of responsible jobs
- Discriminate against minorities, gays, Catholics, etc.
- Dump anyone who is over 40 (or maybe 35), or who is sick

In some areas, the prevalent business culture even restricts personal activities such as dancing and drinking alcoholic beverages (sometimes even coffee). It is not unknown for a company to have on-the-payroll security people follow suspect employees home, especially if there is any possibility that coworkers are sleeping with one another. The justification might be the preservation of the company image or industrial or national security. The telling aspect is that tens of thousands of employees regard these practices as normal, proper, and totally in line with protection of a majority lifestyle. The true result is that the companies cut themselves off from a good chunk of the talent in their industry.

I like to think that we are idealists at heart, even if we pretend to be cynical about many things. We do believe in our work as its own reward, in our employers as gifting us with the chance to do such good stuff, and in the built-in egalitarianism of technology: it's blind to sex, color, height, and social condition. We believe in fairness.

However, the primary function of your employer is to make money, and how it performs depends on the vision and personalities of its founders, the board, a few notably successful managers, and sometimes investors and business consultants. The working recipe or culture may give you an advantage, allowing you access to resources, or it may keep your world-beater concept in the can forever. For example, one company's often expressed reluctance to enter any part of the marketplace where the company could not dictate business ethics and practices. This seemed to be a purely cultural attitude, and it kept the company from effectively marketing or selling their consumer products through normal means.

This chapter gives you a new look at organizations in general, and your organization in particular, and helps you understand its rules and structure. When you structure your own group or project, you will try to form an entity that will work effectively within the larger one and that will have few conflicts.

## THE ORGANIZATIONAL STRUCTURE

A generation ago, a company seemed to consist of little rectangles connected with lines. At the top, the rectangle was named "board of directors," and it was connected to the next-lower box, called "CEO" or maybe "President." Below lay a treelike structure of levels of boxes and branching lines. Somewhere at or beyond the bottom of this dendritic business would be your job.

This box-and-line chart is the classical organizational chart, and it primarily lets everyone know who reports to whom. Sometimes it also, by putting labels against each name, displays how the responsibilities are divided up, and where corporate resources are located. In fancier forms, the chart sprouts dotted lines to indicate secondary authorities and shared resources. The chart displays *responsibilities and authorities*, and as you will see, they don't often wind up congruent like this.

In fact, the chart is a restrictive, low-dimensional way of describing a living entity that has all sorts of ill-defined interrelationships that are important to getting work done. The only case in which a chart of *reporting relationships* is crucial is when the primary company product is reports! (Many wags have defined both the government and megacompanies as closed-loop manufacturers of their own bureaucracy. In these cases the organizational chart actually *is* the product.)

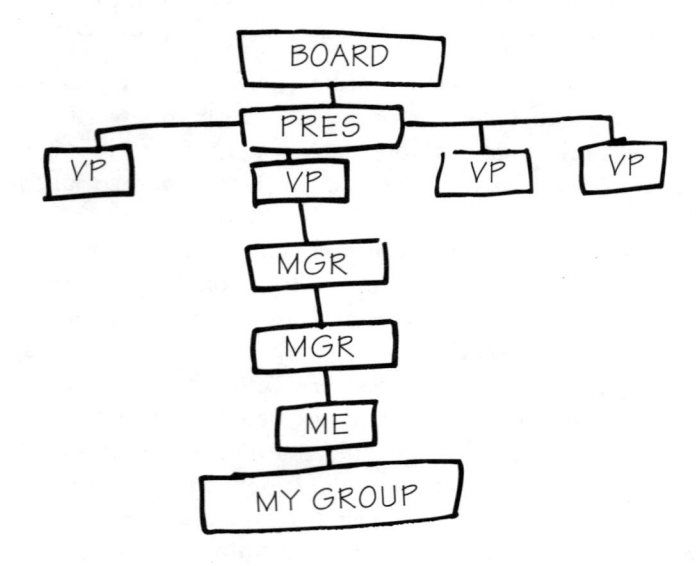

```
                    BOARD
                      |
                    PRES
         _____/ | _____
        |          VP |            |
       VP            |            VP    VP
                    MGR
                     |
                    MGR
                     |
                    ME
                     |
                  MY GROUP
```

In a real company, what you want to see for organizational structure may include:

- Authorities
- Responsibilities
- Resources
- Product flow
- External connections
- And especially: How work gets done

For example, you are somewhere in a chain of events that results in a product going out the door. If you set out to map the product flow from genesis to shipping, you will see that the starting point is somebody, not the CEO, who has a bright idea, and that subsequent events jump over layers of management and nominal areas of responsibility. In fact, it's certain that early on, decisions are made by customers, consultants, and others not necessarily part of the company. The free interchange of information inside and outside the company is vital to competitive products, and this interchange doesn't look like the organizational chart at all. In fact, the deeper you are in high tech, the more erratic product development looks, as each new product follows a different path.

So there arose a form of chart that displays *functions* instead of *authorities*. This gives you a better idea of where to go to get something done, but it still doesn't guide you in finding out *how* to get it done.

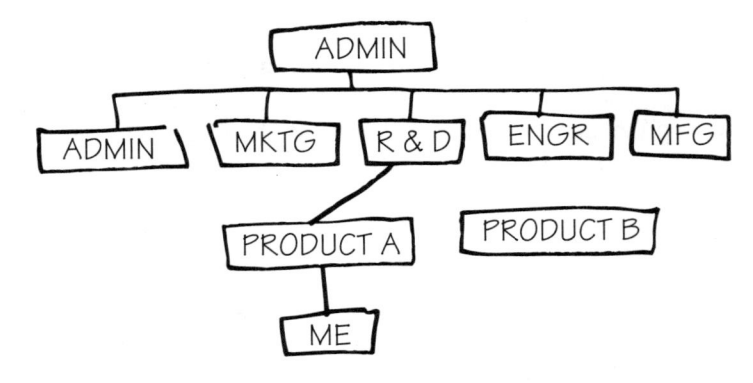

```
                    ADMIN
        ┌─────┬───────┼───────┬─────┐
      ADMIN  MKTG   R & D   ENGR   MFG
                    ╱   ╲
            PRODUCT A   PRODUCT B
                │
                ME
```

♦   The chart you need for getting around inside your organization is invisible and multidimensional, and it develops in your head from experience and continuous experimentation.

## HOW TO DO IT

If it isn't perfectly obvious yet, mull over the concept that, in a high-tech company, the achievers bring in more than their share of the profits. If an employee's time has to return $1.10 for each dollar spent to make it worth keeping that employee (for a hypothetical 10-percent profit), you can bet that some people are returning 100 or 1000:1. Also, some are not worth their dollar. You need to know who the achievers are, both in your own area and in the rest of the company. You also need to know what they actually do. If, for example, you need a quick but accurate opinion on the manufacturability of a new kind of sensor, find out who's got a good batting average on assessments. If you need test equipment, you need to find the person for whom administrative logjams melt. If you need a new idea, find the idea source.

What you're doing here, by asking around and experimenting, is creating that unwritten chart of the actual organization. It's a map of "who really gets what done."

♦   Next, you need to chart the real distribution of authority. This you do with a series of painless experiments. The principle is that you ask many people for the same thing, a sign-off for something, for example. You will immediately find out who approves and who passes the buck. What this means for the future is that you just wave paperwork at the buck-passers, but you personally hand it to the actual one in authority. Note that this chart also doesn't look like the official one.

Now, you're ready to do the same charting for your own group. Where do the ideas come from? Who's good at getting equipment? Who can be let loose in public, or at a customer's? Who's the best engineer? Who doesn't do much of anything?

Because your own group structure is your own responsibility, the point of this last exercise is to try to set up your group's formal structure as close as possible to the way it will actually function. If you succeed, fewer people will get frustrated by having an *authority-responsibility mismatch*, and everything will go more smoothly. Never succumb to the temptation to use the group non-achiever as a buffer between you and everyone else.

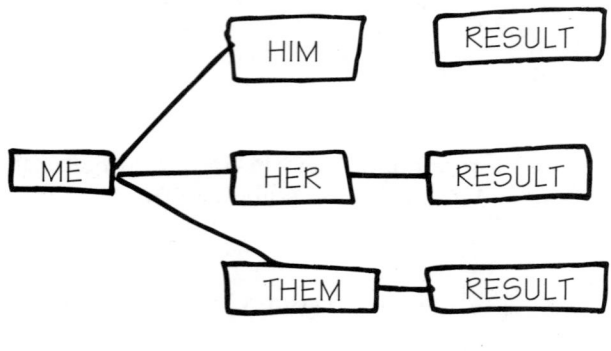

THE ACTUAL CHART

## Vilfredo Pareto and the Support Staff

One way to express Pareto's rule is that the top 20 percent of any company accounts for 80 percent of the profit. Although this is not literally true, it is safe to say that the profit for a high-tech business comes from a small group of innovating, problem-solving brain-workers. Without this group, the product goes stale almost immediately, and the best efforts of manufacturing and sales will not be enough to keep the doors open.

When you get to the chapter on quality, you'll see that it is extremely difficult to measure achievement in our kind of business. The best you can do is to correlate profits with the kinds of work being done. It may be that your company derives no profit from manufacturing your product. For example, some companies sell semiconductor chips that are contracted out to fabricators. These companies have decided that their profitability lies in design and distribution, not in struggling with the tremendous burden of a chip factory.

What this means for a first-level supervisor is that it is not a given that your company has to own a large support, overhead, or even manufacturing staff. You are free to consider what makes the most sense and returns the most profit. This may mean going without a secretary, a model shop, some administrators, or other resources you may have become accustomed to.  ♦  You should concentrate your attention and resources on those people who are worth most to your company.

## YOUR TIGER TEAM

Suppose your bosses thought so much of you and your group that they were willing to give you the most important project, almost unlimited resources, complete freedom to plan your approach, and the promise of some serious reward when you succeeded. If you like this idea, you're ready for the Tiger Team experience.

Whether it is called a Tiger Team, Skunkworks, or the more prosaic Task Force, there is a very special glamour attached to an operation that exists independently of the organizational chart. The idea of an unfettered small group brainstorming their way to extravagant achievement under the very trunk of some ponderous, elephantine outfit has got to be one of the most attractive daydreams in high tech. (The best fantasy, of course, is doing it all yourself!)

Now it happens that few companies, are willing to sponsor a true Skunkworks, and rightly so. The alligators:

- It disrupts and demoralizes everyone else.
- It makes progress and budget less plannable.
- It costs money.

In the next chapter, you'll see how important planning is, and the Tiger Team approach will look like an especially poor bet . . . which it mostly is.

*On the other hand, the most significant advances in your own field may have been done by either individuals or de facto Skunkworks. Over-planned and orchestrated large-scale efforts to develop revolutionary products often collapse under their own weight and inertia. This popular impression is actually true: attempts to force breakthroughs seem to fail.*

What's the problem? The blame often can be laid on irrelevant planning. Plans and schedules are most useful when you've done something similar before successfully. They are less useful when you have to factor in processes involving innovation. Routine work is predictable; pioneering work isn't. This is the way managers have been taught to think. Therefore, a large-scale, multi-million-dollar development program will have all kinds of fine-grained planning detail attached to it but will gloss over or ignore the most important key elements that don't exist yet!

What's wrong with this analysis is that innovation is not completely beyond the reach of management . . . provided that all the characteristics of chapter 2 are in place. When the good idea happens, the organization has to be smart enough to see its significance, flexible enough to do something about it quickly, and upper-level management has to be willing to allow immediate and serious changes of course. In this sense, *any* good group in a high-tech company has to be partly Tiger Team both in attitude and in fact.

The concept of the Tiger Team can be used to motivate and excite your group even though you are all squarely screwed into the normal organization. If you get good at the management job you can, and should, foster the image that everyone's work is vital, advanced, important, and rewarding.

## WHO'S ON THE TEAM?

You have probably inherited most of your group from a prior manager or project. In this group are skills, attitudes, and experience that could be a good fit to the job at hand but that more than likely are not. If you're a new manager, you may have an excessive share of other managers' deadwood, because deadwood can best be removed by transfer to a new manager.

No matter whom you've got, it's a real challenge to define the proper staff for a job, let alone to figure out how the present crew matches up. It's worth noting that, once in a while, someone else's deadwood can become your most valued staff member. Again, if the task has been done before, you know almost exactly whom you need, but if the task is new, you don't.

You start out the same way you start any other planning job, with the final objective. This at least is a given. If you know nothing else, you know what it is you're supposed to deliver. Using the objective, work backward through tentative milestones or achievements and make educated guesses about:

- What distribution of skills is needed
- What other skills might be useful
- What to do if you're wrong about the above
- How much work
- How long do you have

Divide time by rate of work and get people. Write down a list of job titles, how many of each, and total hours per person. Now here's an alligator: If the job has never been done before, you don't know what people you need or how long it will take.

Because I'm a physicist, I used to be convinced that the best choice for any unknown task is, of course, a physicist. The person who can get at any problem from first principles would be the safe bet. Lately, I'm leaning toward farmers, mostly because they have more common sense than physicists. But this is personal prejudice.

Here are a few hints that may help you connect the job to the person:
- Avoid narrow specialists unless you're very sure.
- All experience is relevant, but so is talent.
- Check out your competitor's staff.
- Ask your group for opinions.
- Advanced degrees are rarely important.

**Mushy Marching Orders — The Vague Objective**

Suppose you've been given a nonspecific, mushy objective — for example, "Starting a group to look into new technology applicable to portable workstations." This isn't freedom to choose your own work, this is just your boss not knowing what to ask for. No company pays for completely undirected work. Usually, however, you can dig out some implied or related objectives, such as "this group will also serve as an engineering resource for the California divisions" or "this group will need to identify and evaluate at least one potentially commercial idea every month."

♦ You should be able to form a list of stated and implied objectives and add to it any other goals that you suspect would mesh. Then try to link skills with goals.

In this example, new technology might be optical, xerographic, electro-optical, or other. Do you think of skills in theoretical optics, microfabrication, instrumentation, or what? If you're going to change direction four times a year, you certainly don't want anyone who is absolutely tied to a single discipline, so you're better off going for generalists.

## MATCH SKILLS WITH TASKS

When you don't know which skills go with the project, you've got to do some humbling things, such as asking other managers, copying another group's composition, or seeking the advice of professionals. New managers feel awkward in this situation, because it seems to indicate that they don't really know what they're doing. Get over this barrier; even the most experienced managers are sometimes blind when hiring into technical slots.

Copying another group or another company is a perfectly valid way to start out. If your field involves publication of results, take the most significant recent work on the closest subject, and *see who wrote it*. You'll either know or will be able to find out what backgrounds these authors have, and this information might even help you spot some omissions in your guesses about other skills that are needed.

Getting this particular chunk of data also serves another purpose: When you start hiring, you may go after these same people, get recommendations from them, or get a feeling about which company or university appears to breed the right candidates.

Here are some other places to investigate:

- Professional and trade societies
- Industrial liaison offices of universities
- Consultants
- The telephone book

Of course, you can still be way off base after doing this, but there's no real worry, because:

> *The doer of a task is the only one who can recognize it.*
> Stein's Rules, #422

Four thousand years ago, the guy who came across a hammer, picked it up, and started whacking stuff with it was the same guy who became the carpenter. The other guy, who picked up the pen, just happened to become the scribe. Nothing has changed. Put your task out in the open, and it will be recognized by people who can do it, want to do it, or both.

If you can advertise the objective — that is, put it out in the open — somebody will come along who knows what's involved in doing it or who knows somebody else who does. Magic. This isn't recommended as the sole recruiting technique, of course, but when you have bombed out with other methods, give it a try.

### Human Resources?

Let's be blunt. Your impression of the people in Personnel is not entirely favorable. Then again, until quite recently, you had a limited appreciation of what *any* overhead staff was worth.

However, don't try to save time, money, and grief by making an end run around the Personnel department. While it's tempting, you can be criticized for it, and you are making a long-term mistake. Your managerial talent is largely based on having your own work multiplied and augmented by others, and any time you don't use an available resource, you're missing an opportunity. Cultivate the Human Resources gang, and who knows, you might develop a good working relationship. At the very least, you may help them understand your work and your needs.

## TURNING OVER THE ROCKS

Finding the right people, once you know what skills are needed, is a job involving detective work, team-building, and some understanding of personalities and how groups work. The detective work is in locating prospects, bringing them in, and interviewing them. The rest of the process involves figuring out how these people will work with the others you already have or are going to hire. If anything, assembling an effective group is more difficult than just finding specialists for tasks. It is not something you can pick up overnight. The experienced, successful manager believes this skill works "by feel only," and that some people have the skill and others don't. On the other hand, a large number of managers don't even believe that there is such a thing as group dynamics — everyone does an essentially independent job.

## WHICH ROCKS TO TURN OVER

When you're looking for personnel, you have two general places to look:

- Inside your organization
- Outside: companies, schools, government, independents

Normally, in a big, older company, you *have* to look inside first to fill a position. Something like union rules. The good aspects of this situation are that it's a cheap search, you can help preserve a job, and you can get lots of accurate data on past performance. The bad aspects are that you can be forced to take someone else's reject, you might not be able to get the precise set of skills you need, and you get in-breeding. You'd like, of course, to get "technology transfer" and cross-pollination by hiring from another firm, and you may miss out. The compromise is to hire where and when you can, but to try for the best.

Outside the organization is a large place, unless you work for one of several companies whose culture implies that there is absolutely nothing of value beyond the gates. The methods for hiring outside include word of mouth, advertising, head-hunters, campus recruiting, raiding, and so forth. Because any method can work, the object is to use the most efficient one for your circumstances. Some of these methods will be discussed below, but the guiding principle should be the following:

*The more direct and personal your contact is with the candidate, the lower the error rate.*

## Head-hunters

*Head-hunters* is the familiar term for personnel recruitment professionals or recruiters. They may also be called executive search consultants or search firms. The important things to know are that they are often specialized and have good contacts in exactly the right places. They also can cost from 20 to 50 percent of your hire's annual salary. On the plus side, the best ones will understand your needs surprisingly well and have tremendous motivation to find a good fit, because they don't get paid for people who don't work out, and they do get paid a great deal for those who fit. On the minus side, some companies forbid their use, and they can stretch lots of ethical limits while searching. Make sure that your company is willing to use them. I can't give you any tips on evaluating these firms for fitness to your needs, other than word of mouth. If your field is restrictive, there may be only a few specialist firms available. Otherwise, you have to take your chances. Remember, however, that in a sense the recruiter is representing your company and is making impressions on many potential employees. A callous head-hunter can make a person feel like a commodity, and this does damage.

## Professional Societies

Most professional societies maintain a job exchange, but these are weak and underutilized, mostly because societies cannot afford to manage or advertise the service. Typically, an exchange consists of a free listing in a society publication or on a database. These listings, because they are quite public, are largely composed of people who are currently unemployed, rather than people who are looking to change jobs. Because the jobless are lepers in our world, even though more and more of us are jobless these days, many professional recruiters ignore them. The advantage for you is that you can perform a specific, free search in your targeted disciplines, with a high level of certainty that you can find people who are actually looking for work.

## Your Old School Tie

You may be allowed to recruit on campus, especially if you are a graduate. Companies, large and small, often have a favorite school or two, and it's fun to return and try to explain to an intense young person why you're doing the best job in the world, for the best company in the world. Lacking real-world experience, the new graduate is more than likely to tell you why your company isn't all that good. A few of these interviews can constitute a real Zen adventure, leaving you stunned.

### The Job Bank

Your company usually is required by law to submit all openings to the state unemployment office, which maintains a database. This theoretically nifty concept is corrupted by standard corporate practice, which is to word these job descriptions in a way that would exclude any living human being, again because the unemployed are lepers. If discrimination bothers you, do what you can to use the system properly, and try to change how your company does business.

### The Raid

High tech is given to personnel raids. It's a guaranteed way to get exactly the right staff and eliminate your competitor at the same time. It seems to fit in with diminished corporate loyalties, more mobility, and the incentives of transient employment. You promise more cash, a better jogging track, an office with real walls, and access to flashier hardware, and suddenly your deadliest competitor becomes your employee — together with, it's possible, proprietary information and trade secrets. Lately, because of the legal problem, raiding has become more risky to the raider. Low-level engineering hires are being used to press high-level lawsuits. The line between healthy cross-pollination and theft is being crossed every day. If you want to raid somebody, *please* get cleared by your boss and the legal staff first, preferably in writing.

### Dr. Hammond's Degree

There is a well-known character in the Boston area, we'll call him Hammond, who has held high-level positions with a number of technology firms, but whose entire résumé is a fabrication, including college degrees. He is so well known that people point him out at trade shows and whisper to anybody near, "Do you know about Hammond?"

I'm fascinated by this guy. He moves easily from one company to the next, staying only long enough to avoid being tagged by the inevitable result of his lack of qualification. Each company that he leaves makes no effort to warn the competitor he goes to next, for obvious reasons. He earns a lot of money and gets responsible positions. He's not unique. Someone like him could be your boss.

Professionals have trouble with the concept that unqualified people can talk their way into management jobs. Technical professionals who are managers are particularly riled by the Hammonds among us. We might even be jealous. What hurts the most is that, more often than not, these frauds can make such a good impression that the company is actually pleased with their performance.

On the other hand, some of the most successful people around initially put their hands on rocks and appointed themselves the world experts at this or that . . . and so became.

The moral of the story, if there is one, is that you don't need any specific qualifications — or even any at all — to manage high-tech operations. That is, in this country. Off-shore, it's a different story, and you know the ending.

## INTERVIEW: ART OR FRAUD?

Let's say that you find someone who looks good on paper and sounds intelligent on the phone. The next step is called "the interview," and it's one of the weirdest things you will ever do.

It used to be a meeting between an employer and a prospective employee that allowed both to question and evaluate each other. No more. Now it's an "art," and all of us have had to learn our lines, study videotape, read the law, and generally turn it into a nearly useless, expensive exercise. To begin with, your candidate may well have been coached in body language, appearance, and etiquette, which may disguise important characteristics. In addition, you suffer from not being allowed to ask candidates to demonstrate an important, relevant skill, such as solving an equation or making a computer hum. If you give any sort of a test, you're risking a lawsuit. If you inadvertently ask questions as seemingly innocent as the ages of a candidate's children, you're risking a lawsuit.

Beyond this wall of strictures, you also may not have any time to learn anything crucial about the candidate, because if they've been doing their homework, you'll be doing 90 percent of the talking, and what with Human Resources, the Tour, and others on the interview list (probably more than a dozen), not to mention a lunch or dinner, you could easily spend a few thousand bucks to fly someone across the country for five minutes of significant, one-on-one conversation!

Although there's no space here to cover interviewing techniques, I'd suggest that you look at one of the recommended books or ask your Human Resources people for a book or pamphlet. One of the best bits of advice, seldom followed, is that a hiring manager should write out a script of an interview and have someone experienced look it over.

Interviewing is a learned skill, one of the many we missed on the way to becoming managers. Our tendency is to proselytize the interviewee and sell him or her on the job. This is wrong. The object is to listen rather than talk, to test rather than to show off.

Points to remember: Use interviews sparingly, and get Human Resource's latest update on things not to say or do.

### An Interview with Satan

"You can't be serious!" a coworker says before Dan's interview with a prospective employee. "Elaine doesn't belong here, Dan. I used to work at Dudtek with her, so this isn't gossip. She's a coke head, and she'll literally do anything for drugs."

The law being what it is, there's no way Dan can, even nicely, ask about drug usage. He's put into the uncomfortable position of having to rely on second-hand, possibly erroneous, information. The slightest reference to nonjob matters, and a lawsuit is possible, either now or later, if he hires the person. Dan thinks that Elaine has the background needed for the project, but the tight schedule means that any real unreliability will torpedo the job. On the other hand, discriminating against someone with a health problem is wrong.

The interview goes well, if awkwardly. Dan keeps emphasizing the tight schedule, how understaffed he is, and the necessity of everyone being there full-time and beyond. Elaine says that she can do it; she's a very energetic person. "Energy or chemicals?" wonders Dan to himself. He eventually decides not to hire her.

Quite a few of us are so aberrant in personal habits and behavior that we could be mistaken for drug abusers, the mentally ill, or even the mentally deficient. The best worker I ever hired gave every appearance of substance abuse . . . he was always writhing, twitching, or pacing. But that was just the way he was put together. Another one *did* have a drug problem; he interspersed very good work with periods of unavailability.

You will hire people with problems of all sorts, and you will learn to adapt the work to them. You will also learn something about yourself in how far you can stretch to accommodate the weird, the lame, or the brilliant.

### A Radical Idea

If I were Emperor, know what I'd do? I'd let every high-tech employer interview by try-out. Give your candidate a two-or-three-day shot at *actually doing the work*! This radical idea is actually the way things used to be, in the bad old days. "Here's a drawing board, kid — show me an isometric."

Back then, you could actually fire people at any time if they didn't work out. Now, you have to live with your errors, sometimes for long enough to wipe out your project. Wouldn't it be nice if you could make an interview count?

Well, you can, but you have to be a little devious. Suppose you "show" your candidate a typical engineering meeting. Toss a few questions out. Will you get volunteered opinion? Does he or she shrink from public participation? Suppose you have a sudden problem during the interview day, preferably while the candidate is in the office. What to do? Draw the candidate out; break loose from his or her script.

## THE NEW CHALLENGE OF DIVERSITY

The people who write most new management books have made a fundamental discovery: The work force is diverse and getting more diverse by the minute! This is something like a rural hermit taking a day tour and discovering that New York City has a lot of people in it!

What is this new emphasis on diversity, anyway? It's recognition that not everyone is white, male, and exactly the same age. It's acknowledging that not everyone speaks English or went to the same schools. It's the big news that there's more than one culture in the universe.

It is true that engineering, in particular, has been disproportionally populated by white, English-speaking males having dismally similar educational and cultural backgrounds. It's also true that despite great advances in many areas, engineering has, like a thorny plant, made itself unattractive to women and minorities. Some of the reasons are historical, some relate to American culture, and the rest are connected with educational opportunities. Only a few pioneers will knowingly choose a field in which the chance of promotion is nil or in which an entire career will be spent in isolation.

Then, too, the seeming threat of diversity has caused people to get their backs up and actively oppose the entrance of foreigners and other aliens to their professions. Almost every engineering trade paper has at least one letter to the editor suggesting that job preference be given to native-born Americans (but *not* native Americans). The sciences are bad bets for women in some areas, good bets in others. The government's influence, through its role as a military customer, has been to replace some discriminatory policies with others.

Some of the sins connected with discrimination against people with "diverse" backgrounds are covered by law: gender, race (ethnicity), country of origin, sexual preference. Others, equally important to successful business, are not yet a matter for legislation: age (too young as well as too old), excessive experience, state of health, and ability to withstand physical, chemical, or other work-related abuse.

All the signs are pointing to an impending blow-up in the ranks of the professionally employed over these issues. This blow-up won't be the result of governmental meddling or the exaggerated effects of a few activists — it will be fundamental and will involve all of us. How can this be, when America has always excelled in mixing diverse working groups? The reason is opportunity, or the lack of it. There's real competition for world leadership in manufacturing and high tech, much less fat, and maybe less optimism than ever before.

These factors add up to lower salary growth, fewer chances for new enterprise, less risk-taking, and a degree of managerial hard-nosedness never before seen. Why else would professionals be looking at "union-like" organizations for self-preservation? Why else would the universities see increasing criticism for educating foreigners? Why else does high tech lose ever-increasing numbers of over-30 workers?

What does this mean to you, as a new manager?

To get anything done, you need competent, reliable staff. If you exclude talent because your company won't hire "that kind" or because you can't afford to pay senior-level salaries, or for any other reason, you reduce your chances of getting the job done. When you are forced to have only white, male, single-degreed, 25-to-35-year-old, perfectly healthy employees, you lose. When salary compression forces anyone out of your profession, you lose.

Let's take a closer look at some of the issues.

**Gender**

There are few women in high tech because there's no place to go. There's no place to go because the men have always kept women out of higher-level jobs. The rationale is that women can't deal with predominately male suppliers, contractors, and customers, and besides, the supply of women technologists is limited.

The methods used to exclude women are straightforward: First, restrict the supply by directing no effort to high schools or (nontechnical) colleges to promote opportunities. Second, discourage the moderately interested by not rewarding any women already in the organization. Third, play on American culture by representing jobs as dirty, dangerous, and isolated. (For some reason, however, chemistry and the biological sciences have good representation despite the obvious presence of dirt and danger.) As a result of this history, women stay away unless they either have overriding motivation or know of a specific employer who isn't discriminatory. Then, too, the recent focus on workplace hazards to women of child-bearing age tends to steer them away from hands-on contact with the mystery materials we sometimes use. The overall situation is wasteful and wrong. In the least significant terms, your company is shooting itself in the foot by excluding women in any way. In the most significant terms, it is shooting everybody in the foot.

What can you do? Encourage all the students you encounter to study technology, stressing the fact that it can take less than a generation to change the world. Try to make it possible for women to survive in your group. This means thinking twice before labeling a task as "for men only" or making any job deliberately difficult for women (or anyone)! Conversely, don't perpetuate "women's jobs," such as assembly, process chemistry, and inspection. Whenever you gender-label a task by assuming that "it's too finicky for a man," or "too rough for a woman," you're perpetuating nonsense. If you are a woman and have fought your way into a good career, others should benefit from your experience; hold out a hand.

## Us and Them

For more than 25 years, engineering schools have been attended by large numbers of foreign-born students. The contrast between the population being educated and the population in the workplace is extreme. Either the graduates are going home to use their skills or they can't get jobs here and are trashing their educations by working at something else. Both options mean a loss to us, and the latter means a loss to everybody. Some high-tech industries stand out as significant employers of the foreign born, and others stand out as nonemployers. Your contribution can be to utilize this resource, without worrying about whose job is being "displaced." Make it possible for talent to work here, and you make it possible for your technology to stay here.

## Age

Age is the hot topic. As long-term employment, pensions, and company loyalty fade away, it becomes possible to keep labor cost down by hiring only in a narrow band of ages and ejecting those who are senior enough to command higher salaries. The public reasons are that technologists become obsolete when only a few years out of school and that heads-up, intensive projects require the energy of the young. The real reason is that older employees cost more, both in salary and in overhead. Why keep a 40-year-old around when you can hire two fresh bodies for the same money? Why employ someone who has commitments to family when you can have a 24-hour footloose worker who can be sent anywhere, anytime? Why pay medical insurance for anyone who is likely to need it?

Age discrimination used to be restricted to the over-50 crowd, and pulling the pension rug out from under near-retirees was the sport. Listen up: High-tech business has moved the definition of "over the hill" to include anyone over 35, anyone with a family, and anyone whose prior salary has been "too high." How can you tell? Salary compression. Used to be, the older you were, the more you were paid; it was more or less a straight line. Now, maximal salaries are typically only twice those for new graduates, and if you subtract the effect of engineers-turned-managers, even worse than that. The graph starts to flatten out very early. The message is that experience is worth no extra dollars. Yes, there are exceptions. Some businesses want a few moderately gray heads to use in visible positions when it looks good — for example, when dealing with the Far East or Uncle Sam. But the overwhelming majority of employers want young, cheap, recently schooled, unpolluted workers. So what happens? Older engineers get defensive and start talking about unions, professional licenses, and other protective measures. The schools, which have a vested interest in churning out more product, add to the misery by failing to emphasize life-long or continuing education, thereby making technical education have an even shorter half-life so that more graduates are constantly needed.

You can't fix this situation single-handedly, and you probably will become one of the cast-offs yourself before too long, but you *can* avoid making it any worse. One way is to keep your company investing in training and on-the-job education. Rather than looking for new hires with exact and transient specialized knowledge, spend a few bucks on training someone you already trust. Your company will value only those things in which there's an investment. If your boss can't be convinced that education pays off, everybody's candle is getting short.

You can also help by not creating tasks that discriminate against those who, because of age, family, or health, can't pull 72-hour nonstop efforts twice a week, or who can't grab a toolkit and jump the next plane to Far Away. Although it's true that heroic effort is contagious and results in high output, it's also true that it causes burn-out, as well as lots of accidents and mistakes. At TI Central Research, it was "known" that nobody in the fast track ever took vacation time, and that the often-full parking lot at 3 a.m. was a sign of a well-motivated work force. The company was profiting mightily from this attitude until some of the glamour faded, and the endless supply of brilliant martyrs started to dry up. Not everyone, it seems, is willing to get divorced for the company anymore.

### Health

It's no news that healthcare is failing in this country and that employers have to pay more and more for less and less. What *is* news is that health has become a major source of employment discrimination. The insurer demands that the employer have no risky employees . . . or else. The employer complies by not hiring the lame, the halt, the fat, the pregnant, the smoker, or the freckled. You, as a manager, are cut off from a wide variety of people with possibly incomparable talent. You also can be pressured to fire good people because some bean-counting creep wants to pare insurance costs.

Start-ups and small companies are hurting the worst. They have no leverage over their insurers and must obey. They also tend to have no corporate longevity, so prospective employees who have health problems can't afford the risk of giving up a secure plan for a potentially insecure one. Right now, health insurance can amount to one-third of an employee's salary, and the costs related to on-the-job injuries can be open-ended. You know the signs in your local garage that say, "Insurance Regulations Prohibit Nonemployees . . ." and that help keep kibitzers off the backs of the mechanics? Well, we're getting to the point where we'll have signs that say "Insurance Regulations Prohibit Employees." All of this adds up to pure waste and keeps you from doing your job. Try to dig in your heels when health becomes a reason to discriminate.

The health crisis may be overcome with some form of nationally guaranteed minimum health service, not necessarily tied to employment. It seems to work in other countries. Something on this order will be necessary to allow smaller employers to hire the best people and retain those whose only crime is age or ill-health.

### Kafka Was Here

Today's news included a piece that is of interest to new managers, especially all of you with vats of chemicals, megavolt power supplies, and mutant organisms. As of now, you can't ask a prospective employee if he or she is nuts, or has been nuts. It's discriminatory!

Actually it probably is discriminatory, as are other health-related questions. And mental illness is definitely an illness — sometimes treatable, sometimes not. The ruling, though, forms the finial decoration on a particularly motley structure of law that protects minorities, women, handicapped, and ill workers from discrimination. Like unenforced safety laws, the discrimination laws are given lip service by large organizations and are feared by smaller ones.

In our business, no overt discrimination is necessary, because there are few professional jobs without exclusive qualifications. The few qualified women, minorities, and handicapped applicants who come along are hired without much fuss, because their PR worth is extensive. So, although few of us are sexist or racist, our companies tend toward a white, male homogeneity in the professional slots and a female and minority composition in the hourly departments. However, high-tech professionals don't have licenses or training certifications as do professionals in other fields, such as doctors or plumbers, and anyone can find themselves doing any sort of job, at the employer's discretion.

I have to admit I'd like to discriminate just a wee bit, before I let someone unqualified, untrained, or disturbed near that pressure cylinder of arsine.

## ALLIGATORS

Organizing and staffing alligators are thrashing around everywhere, and you can easily get bit. Here are the big ones:

- Ignorance of the company culture
- Inability to relate tasks to people
- Carrying over bad baggage from other projects
- Not knowing how to find staff
- Poor interviewing skills
- Unconscious discrimination
- No knowledge of employment law and rules

# CHAPTER 4

# Planning Innovation

## PLANNING

Everyone plans. You can't get through a day without planning, from the sock you pull on in the morning to solving problems at work to figuring out where and what dinner will be. Some of us are so taken with this activity that we make little written or electronic plans for the most trivial activities, and elaborate, quantitative plans for more important tasks. In fact, we do so much planning that we take for granted that we know how to do it and that it's intuitive. We rarely sit down and cogitate about the basics of planning methods: how and why they work. Now that you're a manager, it's different. A manager is paid to plan. Planning is so critical to the job that it has to be studied. Special tools have to be mastered. The quality of the plan has to be measured. Finally, the plan has to be tested in the real world.

Planning for, and around, innovation can be different and mystifying. You're called upon to project the effects of an erratic creative process into the future. You need to be able to measure progress that is partly invisible, intangible. The purpose of this chapter is to show you how to approach the planning of "unplannables."

Planning is a management tool. It varies from a verbal agreement to do something today to a mighty document on how to build a space station in ten years. In all cases, a plan contains the following elements:

- Objectives
- Means
- Measurements

Planning starts with its result, the *objective*, which is why *planning is working backwards*. You begin with an objective — a result — and then figure out how to accomplish it (*means*). Eventually you come up with a starting point and a set of directions or a recipe. You also invent ways of *measuring* progress that you can use once the project is under way. If the task involves the coordination of many people, a plan can become complex. If it involves many uncertainties, such as unknown experimental results, a plan can be uncertain and can have many alternative branches. If the plan absolutely must not fail, it has to be comprehensive and must contain back-ups and escape routes. If it must be done for minimum cost, it will look different than another plan for which cost is not as important.

There is no best way to plan a project. Particular methods may be favored in your organization or may generate data that can be compared easily to other data. The plan may outline one or a few major objectives and leave out almost everything else. Or, it could define and measure everyone's activity on an hour-by-hour basis. The plan may look like prose, a spreadsheet, a collection of graphs, or a set of parameters to be logged.

A plan can be efficient in its use of resources and time, or it can be deliberately wasteful. A wasteful plan might be used, for example, to obtain resources for an unsupported project, to protect jobs, or to generate extra profit. More often than we like, we work on plans whose real objectives are different than the stated ones.

## HERE TO THERE

A management plan is similar to a navigational chart, which shows a course from Here to There. The objective is a place and time called There, and the course shown defines how your group is going to get There. There are intermediate objectives along the way, or *milestones*, and points at which you can make positional fixes to find out how far off course you are.

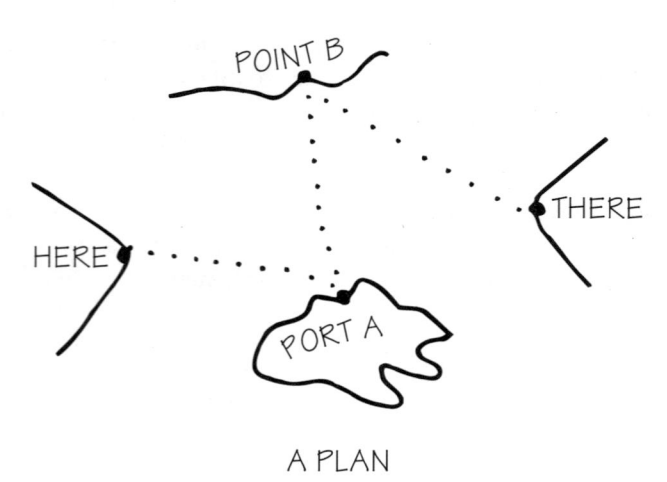

A PLAN

♦ The important and less-than-obvious concept here is that your plan is more useful to you than to anyone else. Your boss is probably only interested in being There; you have to be involved with the details of getting There. He or she just wants it done; your job is figuring out how to get it done. Someone else, a financial person, for example, may only be interested in how much it costs. The technical person doing the work is concerned with pulling an oar but not with steering the boat. The only person who needs to understand and administer the whole plan is you.

This is why I like the analogy of a physical voyage. You're the captain, you're ultimately responsible, and you have the authority.

## DETAILS

Phil manages a software development group. When he first signed on with his company, Phil was unaware that his supervisor had to do any planning. Everyone in the group knew what the job was, and they informally divided the whole tasks into one-person pieces. The job was done when it was done, and there were no noticeable problems. This year, Phil is a supervisor, and he finds that his boss expects to see a quantitative summary of progress, a detailed statement of who is working on what, and a running projection of next week's goals. For the first time, Phil is obliged to consider costs, time limits, and, most puzzling, ways to measure the quality of partially completed work. Although he has a few vague ideas on how to proceed, Phil does the sensible thing and asks another manager at his level how she handles planning and reporting. Then, he meets with his own boss and proposes the same formalism.

Nearly everyone follows Phil's course of action. We find out how someone in our company deals with routine planning, and we copy that technique, if it meets with supervisory approval. We put the system into immediate use and never give it another thought.

If we did engineering or science this way, we'd be open to criticism, and our work would be stagnant. Planning requires education, performance assessment, and continuous improvement. Even when we have to adopt a planning formalism already being used in the company, we should try to figure out why it's good or bad and whether our objectives require changes of approach; we should also constantly be aware of potential improvements.

At present, there are as many planning methods in use as there are different jobs. Although they all have the three elements of objectives, means, and measurements, the formalisms all look different. In the last few years, the sea change has been the availability of spreadsheets, especially those that can produce graphs and charts. Because quantifiable information, such as costs, materials, inventory, labor hours, and sales, are universally tracked with these programs, it is not surprising that they are also used for project planning. In addition, a small number of applications are available that are specific to certain businesses, such as engineering, small contracting, and medicine. You can even download a reasonably useful shareware project-management package. Obviously the future, at least for us.

Your company, because of continuity, culture, or the preferences of individuals, may like to emphasize certain methods and measurements. In manufacturing, the sudden flocking to embrace quality improvement results in much more use of statistical methods for interpreting everything. The techniques of game theory may be more in line with planning in less-certain businesses. Every so often, the Harvard Business School, the Sloan School of Management, and other august academies shift their attention to another area of measurement, and their graduates, as well as their companies, change methodology. Sometimes, individual managers get hung up on single issues, such as variances, to the exclusion of all else.

Even if this were a textbook on planning, there wouldn't be enough room to present the working details of current methods, and much of it would be instantly outdated. For details, look at the suggested reading material and see what's available in your company. My intent is to explain the structure and philosophy of plans, especially as they relate to "unplannable" high-tech endeavors.

## THE BACKWARDNESS OF PLANS — THE OBJECTIVE

All plans start at the end, with an objective, and work backwards. Without a solid objective, you cannot form a strategy to do anything. When you start to work on a plan, you look at the objective and back up toward the initial conditions or starting point, probably before writing anything down. Formal planning then develops from the starting point and works forward.

This first step, examining the objective, is important to you because, in high tech, you may have vague, partial, or speculative objectives. If you worked in low tech, the objectives can be much more concrete: size, shape, cost, time, materials. Learning to plan in low tech really is a matter of optimizing a procedure that has been done before. The planning process then emphasizes the measurement function, as you try to wring out another few percentage points of yield. (I don't want to imply that low tech is easier than high tech, because it isn't, but it is different, especially in this area.)

Your objectives are liable to be lost, changed, surpassed, or ignored during the heat of competition, and planning becomes a much more interesting continuous process of change, redirection, and revision. Instead of "hewing to the line," you've got to keep laying out new lines, changing to new courses, adding and subtracting resources, and figuring out new ways of telling where you are. Sometimes you have the distinct feeling that every time you locate the goal, it gets moved.

Phil, the software development manager, is seeing this course-change characteristic. His group has spent quite a lot of effort writing a package that was intended to run on a specific minicomputer. A quick shift in business partnering, and his company is using another manufacturer's box. Phil can't just trash all the work, he has to adapt what he can and change course, preferably as fast as possible.

At all times, you have to know:

- What you started out to do (objectives)
- What you are doing at present (measurement, strategy)
- What you have already accomplished (results, milestones)

If this looks like potential *spreadsheet* data to you, you're right. If you think that a spreadsheet is comprehensive, you're wrong. In the real world, there are events and developments that *aren't on* your spreadsheet but that change the direction or success of your project. You have the illusion of knowing the detailed status of the job, but you wind up with a canceled project, radically changed specifications, or a missed market anyway. Phil had no reason to plan for, or even suspect, that his company would change hardware. After all, nobody asked for his opinion, and he was directly affected.

To avoid some of these pitfalls, you really need to keep challenging and redefining the objectives, keep working backwards and challenging your strategies, and keep bringing in "outside" data that might be important.

This habit of being open to new directions is easy to acquire if you're a researcher or working at the elusive leading-edge of technology. You already spend quite a bit of time reading, going to conferences, and generally staying on top of developments in your field. ♦ However, you may not be alert to political and business shifts that are just as erratic as scientific discoveries. In the next few years, for example, a vast number of Russian technologists will be trying to shift focus from military to civilian work with no preparation and no help.

---

**Best-laid Plans**

Joe's group is engineering a portable computer. The majority of the design is predictable, but one of the design objectives is to squeeze the maximum battery run time between charges out of a space and weight budget. The chosen strategy is to optimize the combination of nickel-cadmium battery and power supply very carefully so that the least power is wasted in voltage conversion. One consequence of this optimization is that other battery types, such as metal-hydride, will not retrofit. This aspect of the design was locked up early in the planning process. It was locked up because two development efforts, one for each battery type, would have cost too much, and also because Joe was convinced that hydrides were two or three years away from the market.

Joe's equivalent in Osaka, however, took a different tack, so when the hydride batteries did actually become available, Joe was suddenly responsible for making his company's product much less attractive to buyers. He was, however, directly on plan, within budget, and meeting milestones. Do you think that Joe could have avoided this painful experience by putting a contingency in his plan?

---

I'm sure that you know of at least a dozen similar examples of being on track to nowhere. In developmental work, it's really more frequent than success. The most expensive example in my own experience was when I worked on the design of magnetic bubble memories in the late 1970s. At the time, TI, National Semiconductor, and Rockwell, among others, had serious efforts going in the development of these very dense, nonvolatile memories. The science was good, the engineering first rate, and we were just getting megabit chips to work, when the cost and speed advantages of silicon, plus corporate cold feet, abruptly caused *all* the big companies to discontinue work simultaneously. The fact that the product missed its window in the market is not notable; what was much more telling was that the whole industry obeyed herd instinct and packed it in. I'm sure that a first-level manager at Rockwell, say, would never have guessed that loss of competitors would be the worst thing to happen to bubbles!

Here are a few hints for keeping those objectives valid:

- Never suggest an objective just because it's possible.
- An objective isn't good or bad until you evaluate a plan related to it.
- Objectives have to be as detailed as possible.
- Moving targets are okay.
- Reevaluating objectives is part of planning.

Mountain climbers say that they climb because the mountains are there. Technologists like to try to build new things. We have to learn to evaluate new product ideas in terms of market and manufacturability, not only possibility. Nearly anything can be made to work. Not everything will turn a profit.

When you define an objective, you've stated a wish. Until you do some careful planning, including means, resources, and whatever else is important, you can't make much of a statement about the quality of the objective. My inventor, with his time machine, had an objective in mind, but because he had no clear idea of what to do about it, the objective proved to be worthless.

Vague objectives mean vague plans. Detailed and limited objectives can be pursued with accurate and detailed plans. Marketing people are fond of the concept that a new product can be fully defined in the marketing department, and it's nearly true. The better you can specify *where* you're going, the more likely it is that you can pick a good route.

Our objectives change, as we've seen. This is a high-tech fact of life. Because we can't avoid moving targets, we should get used to redefining objectives as necessary. In a progress meeting, you can always bring up the question of the validity of the objective. No part of a plan is carved in stone, not even the goals.

Once you ask the question, you should be prepared to evaluate both the existing objective and any alternative ones. This usually means doing some "what if?" planning. It's a good feeling to go into a meeting with your boss having anticipated a change of direction and having worked up some numbers related to it.

## STRATEGY AND TACTICS

A *strategy* is a choice of means used to meet an objective.

*Tactics* describe the resources used in a strategy.

Joe's strategy in designing the power supply was to optimize the power supply and battery combination, and the associated tactics involved decisions such as which family of devices to use, which technician would build the prototype, and which simulation would be best.

On our voyage, if the objective is a place called There, and we start from Here, one strategy might be to go east to Port A, north to Island B, and northeast to There. The tactics would detail the distances, fuel, weather, tides, resupply points, positional fixes, and so forth.

The distinction between strategy and tactic is fairly subtle, and better terms could be invented, but these are the ones in use. The simple word *plan* might be substituted for *strategy,* and *plan details* for *tactics.* Another way to say it is that a strategy is a general approach, and tactics are the methods used to carry out the strategy.

Just as objectives can shift, making strategies pointless or counterproductive, strategies can shift, making their associated tactics pointless. This also happens a great deal.

If your product encounters a mid-development change, which causes a shift in strategy, beware! The organizational inertia of everyone trying to complete their tactical tasks will run your project right up on the rocks. You, as manager, may be aware of the strategic change, but you have to remember that the people who work for you are focused on their own details and must be *told* to shift their attention.

Changing direction always takes time, from days to years, and all the little plans have to change in order for the big ones to change.

A good strategy can be described as follows:

- It be described in detail.
- It includes no "breakthroughs."
- It is divisible into parts with distinct milestones.
- It is not sacred.

The voyage from Here to There has been described in terms of intermediate destinations. We assume that our ship will go the distance, using the normally prevailing winds and currents. We don't plan on a "breakthrough" such as a strong tailwind all the time or a tow by friendly sea creatures. Arrival at Port A, Island B, and There are good milestones, because we'll be positive that we've achieved each one. If a storm drives us far off course, we'll reconsider the strategy and probably change it.

Changing direction:

- Always has a time constant because of inertia
- Means changing some of the detailed tactics
- Is easiest for a small outfit
- Always causes some waste and damage

The smart, creative members of your group are willing to run hard in a direction that makes sense to them. That's great. However, when you ask them to skid to a stop and go in another direction, they can be slow to abandon good work and get motivated to start over. This is inertia. People who work with ideas tend to get committed to them and have more trouble with redirection than those of us who have little personal stake in the work. Preparing your group to make rapid changes is an important skill, and it is discussed in chapter 5.

When a plan is changed, every tactical detail has to be reevaluated, and possibly changed. It is easy to overlook invisible issues such as previously ordered equipment, facilities requests, or consultant's contracts. Then, when you think that you're on a new tack, some mammoth crate appears on the loading dock with the tool you don't need anymore.

Small companies are often more maneuverable than large ones, mostly because fewer people are involved in approving changes. A five-person start-up can shift its attention in minutes, if necessary. Offsetting this ability is the usual lack of resources. Other groups aren't around to help. Equipment can't be borrowed. Even a new literature search can add delay. Rapid change isn't easy anywhere.

♦ When you drop one strategy and start another one, some fraction of the work performed to date becomes waste, resources have been expended, and the staff has used up some creative energy. If there is too much change, you risk running out of material resources and inducing burn-out. Any control loop can be made too tight, to the detriment of whatever's being controlled. To minimize the damage, you have to develop some insight into how your group functions best, and how its time constant compares to those in the company and in the market. Understanding and coping with change is, all by itself, a substantial subject for further study.

## MILESTONES

Milestones are points in a plan at which you can evaluate progress. They are navigational fixes, intermediate objectives. Usually, they are related to starting or stopping tasks and to easily definable results.

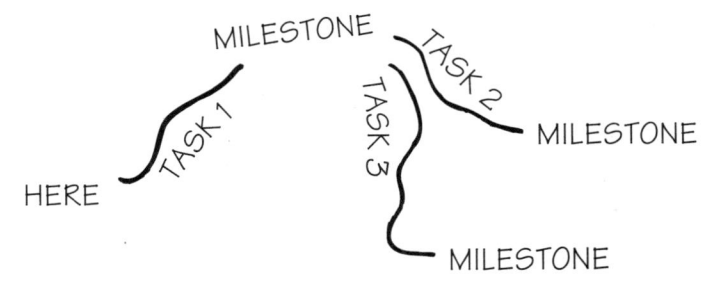

On our voyage, arrival in Port A could be a milestone. It's clear that we're there or we're not, and we also know what date we arrive. If we arrive a day late, according to the plan, we know that we have some kind of deviation and need to consider changing tactics, or even strategy. This is one use for milestones: figuring out whether you're on the plan.

Another use for milestones is as triggers to start or stop other tasks. Demonstration of a working breadboard might be the milestone that both ends the breadboard task and starts the prototype task.

Good milestones have these characteristics:

- They are unambiguous, with clear results.
- They are frequent.
- They might be useful as final objectives by themselves.

For example, "finish some of the work" is a bad milestone, as is "plan review." A good milestone is related to a demontrable achievement, such as "complete layout." Milestones should occur fairly frequently so that they appear always imminent. Every couple of weeks might be average. When a milestone represents a very specific achievement, like designing a subsystem, and the project goes on hold, changes direction, or stops, the milestone can be understood as a final objective. Sometime later on, the design may be picked up for another purpose, becoming the starting point for another plan.

When you use a milestone properly, you compare *where you are* with *where you expected to be*. You then have to do something about the difference. You might have to ask everyone to work faster, change the strategy, or back up and redo a task. The important thing is that the milestone is also a point at which you make decisions, and what you decide to do has to be an option in your plan. Wise old managers say:

> *If you know where you're at, but don't know what to do about it, you haven't achieved much.*

## STRATEGIC PLANNING

*Strategic planning* is a term reserved, in most instances, for the process of picking the top-level direction for the future. This has little or nothing to do with the sort of strategy we're describing here, but you should be aware of the usage.

## RESOURCES

In planning, you use the following resources:

- People
- Money
- Equipment
- Space
- Knowledge

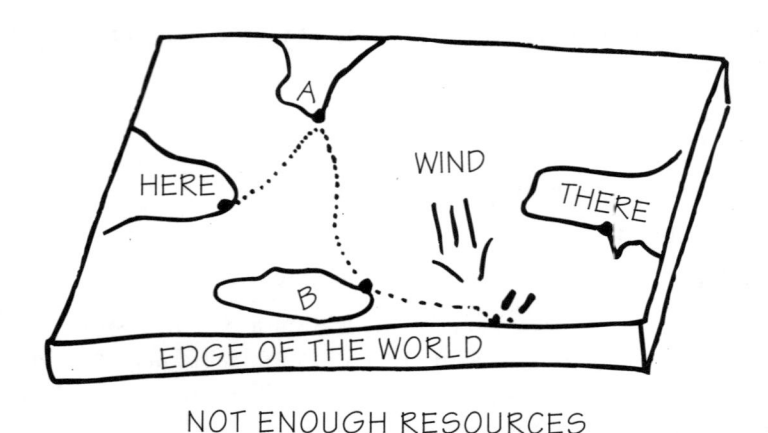

NOT ENOUGH RESOURCES

A resource is something you can control. Time, external forces, decisions by higher-ups, accidents, and other uncontrollables are not resources. Nonetheless, they can wipe out plans and resources. The sum of all the resources could be called *technology*.

Each type of resource enters a plan in a different way. You use up money and equipment but you don't use up your staff (you hope), and nobody has figured out how to use up knowledge. Also, each resource involves actions by other people in the company. Money, for example, is meted out by financial people who have different objectives than yours and who handle dollars in ways that are totally irrelevant to you, possibly even incomprehensible. In fact, dealing with each resource requires specialized knowledge that you cannot be expected to master. Instead, you will learn how to communicate with those who provide resources, so that they can get you what you need. In turn, you act as a provider of resources for your group. When someone needs more space, temporary help, or a piece of equipment, you're the one who handles the request. You can become good at this only if you take the time to learn how the organization functions and where to go for resources.

**Money**

A technical plan is not a budget. You may have to know how much your people cost (that is, how much your company charges your project for each employee), how much everything else costs, what's capital equipment, which cost levels require which signatures, and how other groups charge you for their services. (If you don't have project responsibilities, you may not do any real budgeting.) Because each expenditure comes from a different pile of money, and since the bean counters may treat each category with special rules, it's unlikely that you'll understand all the subtleties of overhead, capitalization, write-downs, and amounts charged to contracts you never heard of. One of the first steps you should take is to find out who will do the financial work for your group. The person doing the financial work may be your boss, a contract administrator, an accountant, or a financial officer. Make an appointment and get a list of categories that you need to understand.

Then, if your boss tells you that you have $1.0 million to spend, you can start dividing it up according to the categories that are being used in your area. You will put costs in wrong places when you start out, which can be forgiven. What is not forgiven is making the same errors over and over again. You waste your time repeating paperwork, and the financial people waste their time correcting you. If you start out with a good working relationship, ask questions, and allow them to help, budgetary items will be easy to handle. In dealing with financial issues and people, follow these guidelines:

- Ask what each resource costs.
- Use, and reference, *their* numbers.

- Use *their* categories, not your own.
- Ask instead of complaining.

## Equipment

High tech is equipment intensive. You need fancy toys, new toys, and very expensive toys. You may use some expensive tool only once (test equipment comes to mind) and then store it or move it into surplus equipment. You borrow and scrounge stuff, and when you can't avoid it, you build things. The result is that your company has a mixture of stray, home-built, and stripped equipment that circulates without any paperwork and that gets lost, broken, or stolen. Record-keeping is horrible, and the inventory process is usually a joke or a nightmare, depending on whether you have anything to do with it.

Each item you use turns up in your plan differently, according to its type, age, state of depreciation, or origin. Items can be borrowed or leased, and time can be rented. If you're expected to keep track of all this, you wouldn't have time to do anything else. However, if you ignore equipment cost and control, you'll end up not being able to buy what you need or losing money that you could otherwise spend. If you don't pay enough attention, some large expense for something that you don't use will cripple your project — invisibly. Again, the best strategy is to make friends with the financial department and get their help. Then:

- Use their categories on equipment.
- Get rid of things you don't need.
- Find out if borrowed and scrounged stuff is charged to you.
- Be careful with buy/build/lease choices.

Once, when I supervised a process development area at a semiconductor capital equipment company, my primary tools were a couple of the company's latest systems. Because these were treated as mules (that is, they were often modified and could never be sold to a customer), I never assumed that they could be on my books for anything like the $600,000 sticker price ... but they were.

Getting rid of equipment is a touchy subject with some of us. There are packrats in every company. Anything you're keeping is costing you money, even if it's only through floor space. See if you can return items to inventory, to a warehouse, to your department. You can, with appropriate official blessing, even sell excess equipment.

In general, it never pays to build anything you can buy or lease. We may all claim to know this, but most of us look at some trivial bit of equipment with a big price tag and develop an urge to save the company money by building a copy. That's how it happens that a senior-level engineer, never fully grown out of Heathkits, ends up spending a few weeks (at a cost to the company of $4,000 a week) putting together a power supply that could be bought for $1,500.

As usual, if you're in a small company, you may not have as many equipment resources available and purchases may have to be minimal, but you may also have far fewer administrative problems dealing with equipment. In a government agency, for example, the overhead connected with buying something can easily cost more than the item itself, and the process can take so long that the item is obsolete or no longer needed. In the government and in some large companies, the most practical options, such as leasing or borrowing, are often made impossible by administrative rules. If the money is there, and you can't spend it when you need it, you're being held back. In high tech, being held back is fatal.

If you happen to be working in one of these clumsy environments, you can compensate to a degree by searching out the acquisition wizard. There is usually someone who has skill in bypassing all sorts of impedances. Because they accomplish their wizardry by trading or bartering, you need to have something to trade.

## Space

Facilities and space are almost beyond your control, even if they are resources. If your plan depends on getting some plumbing, wiring, or ventilation, take your most reasonable time estimate and triple it. If you've got some money, you can probably buy an oscilloscope, but there is no way you can buy the attention of a facilities manager. Even safety issues won't accelerate anything. I know of a few places where the fume hoods arrived after the first mortalities — and that's a long time. I've never been certain why companies don't structure supporting departments a little better, although I have a few theories.

At any rate, you'll need to avoid getting caught by not getting the space and utilities you need. One way is to discuss before you specify. An extra gas line or drain might be penciled in without any thought on a facilities request, but the seemingly innocent addition could require some disproportionately massive efforts somewhere else that could cost a year's delay. Only you can know what you *don't* need and can afford to give up. Find out what's a problem and what's not. Another ploy is to replace fixtures you can't get with portables you can install yourself. A good example is a three-phase power line. It might take a long time to get a piece of conduit with a receptacle on the end, but it takes no time at all to run an extension cord.

### Knowledge

Knowledge resources are in your staff's minds, other people's minds, libraries, databases, notebooks, cooperatives. These resources are in your plan. You already know where to go for journal articles, and you may know where to find the company experts in relevant technical areas. Do you know when to apply for a patent, and where? Do you know about industrial liaison programs your company has with universities? Do you know if consultants are available?

In writing a plan, you need to have a handle on the availability and quality of knowledge in your company, and outside of it. From lab turnaround to lawyer responsiveness, the success of your project depends on your knowing who else knows what — and how fast they work.

## MAKING A GOOD PLAN

A plan is only a map of a voyage, not the voyage itself. The map can be distorted, have errors, or not extend far enough in the direction you want to go.

For a plan to be useful, it has to model reality with enough accuracy to help predict events in the future. It also has to be simple enough to be understood and manipulated both by its author and by others, and it has to be testable against actual events. Because the "weather" on your voyage is not under your control, your plan has to be comprehensive enough to help you find your way after being blown off course. In our trade, it also has to include the possibility and effects of innovation and frequent failure.

A good plan is *accurate, simple, and complete.*

### Accuracy

In low tech, a production plan may mirror actual events to within a few percentage points. That is, the objectives of output and cost versus time may be met very closely. In high tech, our objectives are chancier, and we may exceed them by factors of hundreds or thousands or miss them entirely. A gravel plant will run at highly predictable cost and yield highly predictable product. A company developing new process catalysts may produce nothing, break even, or hit paydirt, so to speak. The definition of what constitutes an accurate plan varies tremendously.

Accuracy, in a planning sense, means more than having the budget work out properly. The way I see it, accuracy is related to achieving the objectives, and the rest of the plan can be changed, added to, or ignored. If you agree, your plan has to include some way of *measuring* the final product against your initial intent. This gets into the area of quality, which has a chapter to itself.

### Simplicity

Just having a spreadsheet that covers a matrix of thousands of rows and columns doesn't mean that it helps you understand a project better. Just as managers have trouble with more than a handful of direct subordinates, people have no ability to manipulate overcomplex structures in any way except piecemeal, or in simple overviews. I like to restrict a plan to a single, final objective, two or three tasks per person, fewer than ten milestones. If I can't see the main structure of a plan in large type — on an overhead transparency format, for example — it's too complex. Nothing keeps you from making subplans to cover the smaller details and tactics. Just don't clutter up the main one.

### Completeness

Even though the plan has to be simple, it has to show enough "off-course" territory to help when the unexpected breakthrough changes everything. This means putting in some "what-if" branch points and making some educated guesses. It's a really good feeling when you can tell your boss that you have already considered a contingency that actually arose.

To make your plans complete, involve everyone you can in brainstorming the possibilities. You have all these experienced professionals around, and it's just possible that they have seen or can hypothesize many possibilities that you cannot.

On the other hand, a plan can lose focus if you elaborate it with low-probability contingencies. Design of a nuclear power plant, for example, is heavily biased toward the anticipation of low-probability events . . . to such a degree that the project slowly becomes unachievable in terms of the primary objective of so much cost per unit of energy. Remember that the plan isn't a fixed and sacred document. By itself, a plan is not a guarantee of success.

## CRITICAL ELEMENTS OF ANY PLAN

Any plan must have the following:

- A relationship between time and resources
- Statement of objectives
- A way to measure progress

Most plans can be summarized as a chart that shows tasks versus time. Connected to each task are people and other resources. The commonest one is the Gantt chart, which is a list of task descriptions in bar chart form, but there are any number of ways to show similar information. In most formats, a task has a starting and ending point and may be conditionally related to some other task. The ending point is probably a milestone. The rightmost column is the end of the project. (Time seems to go left to right, except for financial purposes, when it goes top to bottom.) Once, to add a little interest to project review meetings, I converted a Gantt chart to look like a horse race, attaching a drawing of a galloping horse to each task's current status. Because all the horses were supposed to line up vertically if the plan was on track, lagging and leading tasks stood out.

What's missing from this kind of graphic display? Most of the significant interactions between different tasks. If we represented these, there would be all sorts of spider webs connecting everything with everything else, and the plan would look silly. However, as long as people don't actually work independently and there is a group synergy, these interactions exist and are important to getting the work done. It's just that representation on a graph isn't easy. The prose part of a plan will contain this information — for example, "Joe will adjust his design as test results are available from Sally."

♦ The "outside" events and forces, which we also omit because we can't control them, might be thought of as events out of the plane of the paper — that is, in the third dimension. In other words, when the research group at Steady State University develops a fabrication breakthrough that comes to your notice on January 7, changing your project's direction, their chart intersects yours on that date. I think of an amoeba ambling along, dividing and replicating according to its chemical plan, when zap! a cosmic ray passes through and alters the plan.

The Gantt chart, or any other formalism, isn't necessarily the best or most complete representation of a plan. You can invent any system that helps you understand and communicate the factors that are most important in your particular work. The best system for your own understanding might be terrible for communicating with others, possibly because the information might not be simply displayed. If you use, and are familiar with, spreadsheets, you can easily extract substance from all the data, especially while you're sitting at the keyboard. The same information, frozen on a presentation graph in a meeting, can't be manipulated or enhanced and therefore won't communicate well.

Objectives have to be stated as completely as you can, for two reasons. The first is that, given enough detail in the objective, making the plan becomes easy. The second is the need to measure how close you have come to meeting the goals. Because we get hit with changing objectives and decide to sail to another destination, everyone involved with doing or approving the work must frequently refer to a good, current statement of objectives to be sure that we're all working on the same thing!

## Measuring Progress

Here again, the difference between low tech and high tech is in how quantitative the measure of achievement can be. Every morning the doormat production crew can evaluate how much they have done in the past 24 hours and how much of that has passed inspection. It's anybody's guess what a manager of theoretical physicists can measure on a daily basis. How do you measure the amount or significance of innovation? How do you figure the yield on invention?

### MEASUREMENT AGAINST PLAN

Measuring progress in research has always been a fuzzy task, but it has become more important lately for justifying expensive work to those who would fund it.

These are some common, but bad, measurements:

- Number of publications
- Money spent
- Labor hours spent
- Company internal "attaboys"

Here are some good measurements:

- Successful hand-offs to production
- Achievement of milestones
- Something the outside world deems interesting
- Clear facilitation of another task

If you look at your company's annual report, you may see some of the bad measurements in use. The company is clearly trying to show the shareholders that it spends a lot on R&D, so the authors of the report add lots of nonresearch money (such as product development cost) and cite achievements that are of little or no interest to the rest of the world. If you've been fighting for upgrading your lab or adding another hire, you can really wonder where exactly all those millions were spent.

In government agencies, the number of publications is often used to measure progress. What could be presented in a single, significant paper gets fractionated into dozens of relatively trivial ones or repeated with slight changes year after year. The volume of paper increases much faster than the usable information content. However, publication volume is one of the few accessible measurements, so it becomes overused.

When you measure progress, you are really trying to see where your plan is going and whether the objectives are going to be met. This is why you need a plan that helps you identify demonstrable achievement. Your milestones are achievements. A usable idea for a new product is an achievement. Transfer of technology or assistance to another group or division counts positively. In the commercial sector, in fact, you measure company progress in dollars of profit. How your group contributes to those dollars is a valid measure of progress.

## PLANNING IN THE INNOVATION BUSINESS

Planning in, for, and around innovation is a serious subject. We can look at giants, such as Bell Labs or Xerox, or at flashy start-ups, such as, most recently, companies in genetic engineering. We may not see any commonality. A great deal of success appears to be luck, timing, and financial competence. A single invention, such as xerography, can come out of unsupported, dedicated work by an individual and give rise to an entire industry. Thousands of carefully managed research projects can result in thousands of technological advances and new products, or they can result in almost nothing. It has been said frequently that there have been no major developments in electronics since the 1967 introduction of the integrated circuit, but there obviously have been quite a few minor ones!

There are two kinds of innovation:

- Foreseeable innovation
- Breakthroughs

In the development of semiconductor integrated circuits, a trend was spotted by Gordon Moore in the early days: Complexity (density) doubles every 18 months. In order for this *foreseeable innovation* to happen, tens of thousands of technical developments and inventions had to occur. Many of these developments individually looked like breakthroughs, but none of them were revolutionary enough to alter the large-scale trend.

On the other hand, the invention of the transistor was a *breakthrough* in that it caused a sudden, nonevolutionary change in electronics.

In your business, you have a mixture of both kinds of innovation. The first kind is predictable. You know that there will be a wristwatch telephone not too far from now; all the basic elements (high density, low power circuits, a cellular repeater network) are in place to make it possible. The question is when. Will the demand be sufficient by next year, five years from now? In this case, demand for the innovation will lead development and widespread use. Breakthroughs, on the other hand, are erratic, coming from the synthesis of hitherto unrelated data. A single, minor breakthrough might be enough to give you a viable product, or you might need several. It's understood that mechanical, rotating memories represent an obstacle to new products based on microprocessors. A breakthrough facilitating whole new classes of portable products might be a new type of solid state memory — for example, one that uses the volume of a chip rather than just one surface, or optical rather than wire connections.

With breakthroughs, the question is not only when, but what. The futurists of the early part of this century saw us all flying around in personal planes, communicating by television-like means, and building ever-larger cities. The actual breakthroughs were not foreseen, nor were their effects.

Innovation has another characteristic that you must understand: New and important concepts are not always recognized or developed immediately. Sometimes they are ignored, buried in proprietary vaults, or improperly promoted (see below). It can happen that a breakthrough blossoms in your very group but is not noticed. Why should you care? After all, you probably get only one buck as token payment for each patent granted. The reason you should pay attention is related to your being a good manager. Your people must feel that their work is recognized, or they will avoid the treacherous shoals of creativity. It's nearly impossible to do outstanding work if there is any reason to believe that the idea will be ignored. Just one experience of this phenomenon can embitter a good technologist for a lifetime.

## SETTING THE STAGE

If you want to foster innovation, of either kind, you have do develop the correct environment for it. Heads of major research enterprises have written about how this is done in their empires. There seem to be many approaches, ranging from hands-off management to highly focused and closely managed teams. Only a few of us have the luxury of working on whatever seems interesting at the moment, so we have to actively manage the creative process.

A common denominator, however, is *flexibility*, both in planning and in managing. When you make a plan, you're putting limits on what's being worked on, how long, by whom, and with what tools. If these limits are too wide, you're not managing with specific objectives in mind, but if they're too narrow, nobody will have a chance to check out potentially key concepts. You have to be flexible in your approach to organization and planning, so that the novel stuff — which is always a deviation from plan — can develop.

The other common characteristic of a good environment for innovation is *recognition*. If a breakthrough occurs, you have to be able to see it. This is not guaranteed. It's easy to miss the significance of a new idea or to expect that great concepts cannot come from the rank and file. A lowly technician in a laboratory not known for fundamental discovery probably made the first measurement of superconducting tunneling, correctly but tentatively identified it, but was told by his boss that it couldn't be. Thus the data was buried and the Nobel Prize went elsewhere. In another instance, equally frustrating, a colleague of mine could not convince a high-caliber research organization that his new concept for an instrument was worth funding. He also got to see a Nobel go elsewhere . . . to a lab where the potential was recognized. Today, the instrument gives us our best images of the atomic-scale world.

You have to develop an eye for innovation and be able to distinguish between what's actually new and what's just new to you. This isn't made easier by the habit of many organizations of treating wheel reinvention as true innovation. You also have to avoid instant dismissal of new ideas because you doubt that great ideas can come from ordinary people. You can best develop these skills when you're not inhibited by your company's rigidity or your own.

Another necessary component to the innovation environment is making the rewards look good enough. Because there isn't much company loyalty left among employees these days, your Edison has to be convinced that offering up a potentially portable and lucrative idea to the company is better than running off with it. Most of the attempts to reward technical people (dual ladders, fellowships, and so forth) have been failures, and at present, no particularly foolproof ways exist to keep a great concept from migrating. The best companies keep their innovators by respecting their ideas, following through with products or more research, and providing needed resources for new work.

## ON THE PLAN

If you draw up a plan that has unknown results in it, it will look like a tree, as in the sketch of the "isles of development." Each time you get to the unknown part, there will be multiple directions you can go, depending on the result of an experiment or an innovative idea. A perfectly valid outcome of an experiment might be that the project simply fails and ends right there. Or, several new avenues might be started. If you're used to diagramming programs instead of voyages of discovery, think of a program flow chart, complete with conditional points, loops, and branches.

# Planning Innovation

To keep the plan simple, a single objective shouldn't be based on more than two or perhaps three innovations or breakthroughs, because the amount of branching due to these unknowns gets awkward. What you do instead is break up the objective into smaller subobjectives, each with its own plan. All you've done is replace the chart from Here to There with a number of charts to many potential intermediate places.

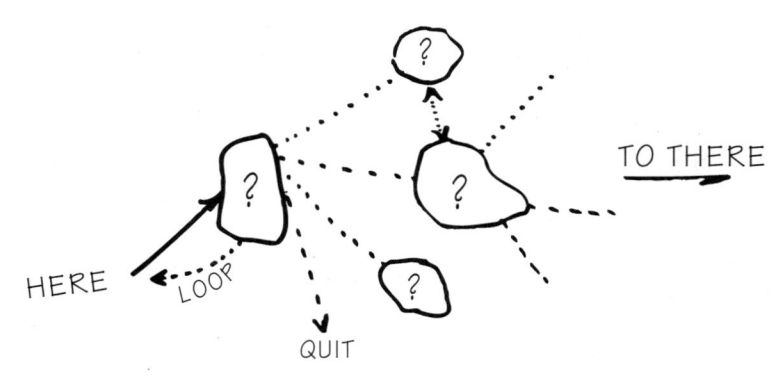

THE ISLES OF DEVELOPMENT

*Now, here's how to relate this sketch to the more familiar plan. Replace each "question mark result" with a milestone, and then make the route up to each milestone a task.*

◆ If you get in the habit of thinking about unpredictable developments, results of experiments, and breakthroughs as milestone points in a plan from which radiate several alternative plans, you can reduce a complex and uncertain developmental project to a small number of simple and completely conventional plans.

◆ To put it another way, you don't know what exactly will happen at each milestone. Instead, you *plan* to evaluate the project status at many decision points and make reasonable new plans based on what actually happens. As you become skilled at this, your ability to change direction becomes enhanced, and you waste less time and fewer resources following the new course.

◆ Are there drawbacks to this technique? Yes, of course there are. The biggest one is that you give up information about how tasks relate to one another, which is the normal failing of most chart-based plans. Remember that all the interconnectivity is not shown. This makes it easier to get lost, to duplicate prior work, and to stray from accomplishing the main, final objective. You can lead your group into a purely *reactive* state and lose continuity, lose time, and get butchered by the competition.

A manager in high tech has to solve the planning-in-uncertainty puzzle over and over again. Part of the solution is defining and never forgetting the objectives. Another part is constructing a plan that incorporates unknowns as routine events. The manager also has to understand and nurture the innovation process, making room for new ideas that "don't fit." Finally, despite plans, fancy managerial software, models, and theories, you have to stay focused on your job as a motivator, provider, and leader of other people. *In the final analysis, you make it possible for innovation to prosper, and for innovators to to do their very best.*

## WHOSE OBJECTIVE? A CASE STUDY

Bernie is an internationally known scientist working in polymer chemistry. He happens to work for a company that produces plastic web materials for packaging and coating. Bernie was the first heavy duty hire for this manufacturer, when the company had grown to the point where a research department with a legitimate head (this means a Ph.D. with publications) was necessary to address new product needs and properly market to customers.

The problem is that Bernie, while not refusing to take his paycheck, or the status of a supervisor, has launched several projects of his own that don't show on the plan. He has become interested in polyamides, which, of course, might eventually become company products but which are not supposedly being considered. This work has been taking more and more time and attention lately, at the expense of more relevant work. The first to complain openly is the marketing crew, because Bernie is too busy to talk to customers, what with publishing, the new apparatus, and conference-going. Bernie's supervisor realizes that he's supporting basic research on a development project, and has mixed feelings about that, but he's actually afraid to confront Bernie. Bernie is, after all, the Grand Pooh-Bah of Polymers.

What actually happened here to resolve this problem was that Bernie was fired, on orders from two levels above. To the executive making that decision, the situation was clear; someone on payroll was doing his own business on company time. That person could be replaced by someone who wouldn't steal from the company. There were no warnings, no bad annual reviews. Bernie was just told to leave by the end of the day.

In your group, you will always have some amount of "off-plan" work in progress. The trouble with R&D is that many of us regard the freedom to pursue our own interests as important — to us, to the company, and to the state of knowledge. What we don't realize is how expensive this is to our employers. An industrial scientist with two technicians can spend more in a month than a counterpart at a university spends in a year. Your job is to keep your Bernies on a leash without stifling creativity or innovation. I hope that you can see a better way to do it than firing people.

# CHAPTER 5

# Working Faster

## SPEEDING UP

You can see it happen. The time you had on a project a few years ago seems like a luxury today. Every advance in technology has the effect of making more work possible in less time. Consider the CAD workstation, automated prototyping, engineering analysis software, and fancier test and analytical gear. All of it helps reduce the time to market, collapses your development time, and accelerates your project. The problem is that it also accelerates the competitors' efforts. Schedules that would have been unthinkably aggressive a short time ago are now necessary if you want to stay ahead of your competition.

The hitch in this high-tech getalong is simple: Not all the tools and techniques advance at the same rate, leaving you with a mix of faster, more efficient processes and slow, potentially crippling old ones. Here's an example: It's great if you can do a circuit board layout in minutes instead of days, but if you lose the time gained to meetings and paperwork, you're losing ground. If it takes you two months of dickering with the company legal crew to buy a day of a consultant's time, it doesn't help that the consultant delivers in two days instead of a week. To work faster, you have to speed *everything* up, because if you don't, somebody else will.

Innovation doesn't have its own unpredictable pace any more. A great idea can't rumble around undeveloped for years, or be investigated on a shoestring, after-hours basis. The competition will eat your lunch. Any viable company has to be willing to go after new product with every available resource, and do it faster than last year.

What results, of course, is that bigger, more expensive mistakes are made, dead ends are followed too far, and the cost of innovation keeps rising faster than the profits. This environment is murderous to start-ups and the Tom Edisons who must be out there somewhere.

Smart management is part of the cure for this problem. You, as a smart manager, have to develop ways of working faster, finding mistakes earlier, responding more effectively to change, and keeping better informed. You have to spot the bottlenecks, the dead ends, and the outdated administrative practices, and you have to do something about them. This, naturally, is much easier said than done. This chapter examines the barriers to working faster and tries to separate the techniques of the hares from those of the tortoises.

## PRODUCT CYCLE TIME

The French Academy thinks that a century is not too long to spend putting together a dictionary. The Mee-Too Toy Company of Brooklyn making knock-off clones of the popular Middle-Aged Presbyopic Couch Potatoes will go from concept to shipped product in a week. Your newspaper funnels information into print in hours. It all depends on competition and the market. Apparent complexity, new technology, wishful thinking — they don't count for much. If the product *has* to be developed in a big hurry, it will be.

Even though you're embedded in the process at a single point — manufacturing engineering, for example — you must understand the way a product goes from demand to satisfaction, or from idea to crates on the dock. The phases of the cycle are usually as follows:

- Real or potential need in the market
- Critical mass of concept or technology
- Recognition of viable concept
- Commitment of resources to development
- Proof of concept, prototype, breadboard, etc.
- Evaluation
- Manufacturing engineering
- Evaluation
- Production
- Satisfaction of market need

Who's responsible for each of these steps? In a tiny start-up, the entire cycle can actually be the work of one person, which at least saves on time wasted at meetings! For larger companies, the tasks belong to specialists and teams, although there can be substantial overlap.

The market need is either dropped on the doorstep by a customer or sensed by marketing personnel. The first route is at least as valid and efficient as the second. Marketing may "discover" a need by projecting current technology into the future, observing related developments that facilitate new product, or copying what a competitor is doing.

Marketing may then ask the R&D department, the chief technologist, or the advanced products group to look into the possible satisfaction of the need in terms of a product. The response may come from a literature search, evaluation of new technology, espionage, or actual research. The form of the response is, "Yes, we have the technology."

A conceptual package is assembled by everyone involved up to this point and is presented to upper management as a proposal. Based on additional knowledge (the company's finances, long range and strategic plans, and entirely personal preferences), a decision is made to pursue the new idea, and an initial plan is constructed. That plan gives an outline of how long, who, and how much. Management then commits resources to development.

Development consists of many stages, such as proof of concept, prototype, breadboard, brassboard, and others unique to certain product types. The work is performed by scientists, engineers, technicians, and a host of supporting services, such as computing, analytical, and fabrication departments. The result of all this may be a physical prototype of a piece of hardware, an abbreviated collection of software, or even a numerical model.

This prototype is then evaluated by customers, engineering, manufacturing, and marketing departments. Bugs are found, features are added or deleted, and the prototype is passed on to the people who have to design the real thing. The hand-off from development to production (or to engineering) is a critical stage, in that management must commit more resources and work the product into higher-level plans, usually with a firm date for its introduction.

In what should be a highly interactive process, engineering and manufacturing personnel work together to come up with a production-worthy design that also can be built, and even serviced. They again have the assistance of the customers, the R&D department, and anyone else with an opinion.

Following more evaluation, production starts, problems are solved, and the first units ship to users. If everything has gone properly and quickly, a market need is satisfied, and everyone is happy.

---

### Product Cycling to Nowhere

It's absolutely amazing how often a company gets all the intermediate steps right but overlooks the first and last step. Things get built that have no market, and things get built that are never accepted or supported.

From luxury cars introduced in a depression to home video recorders in the days before standard formats, manufacturers have forged ahead with products because their timing was off, market conditions changed, or the perception of the market was faulty. Because high-tech products are heavily influenced by the perceptions of technical people, we always have to be alert to the danger of assuming that our customer is as entranced as we are with a new and flashy concept. The assumption that every household would need and use a personal computer, in the early days when they were fairly clumsy devices, was no doubt promoted by engineers. Several companies went under trying to produce factory robots for customers who found that human labor was less expensive.

Although we don't have to be marketing or sales geniuses, we do have a responsibility to ourselves, our subordinates, and our company to understand enough about the need for and end use of the products we build. We can't just do our own small contribution to the process in isolation . . . the way it used to be. Despite the existence of many professionals in the product development cycle, we cannot assume that the others have spotted or understood all the issues, or that our marching orders are flawless.

---

The product cycle time is how long it takes to run through this sequence. Some of the steps are done over and over again — experiments or revisions. Sometimes the sequence takes a sharp turn toward a different concept because of external developments, and sometimes the whole thing is trashed. The bubble memory effort mentioned earlier is a good example.

The planning that goes into a product introduction looks just like the group-level plans you create. Objectives (successful product) are connected with strategies that which specify time limits, costs, and resource levels. All the small plans are scooped up and assembled into master plans. Each of the phases of the product cycle generates clear milestones, such as "demonstration of breadboard." As departmental efforts run early or late, the master plan is modified to suit, and you get feedback that allows you a little breathing space or points toward an all-out crash effort.

## THE COMMON AGENDA

Not everyone in your company appreciates the big picture. In fact, there is no way that every manager can understand the dynamics of the market and the dynamics of the company. It's rough enough just handling one group, one objective. However, if all the oars aren't simultaneously in the water and pulling, nobody goes anywhere. What this means is that one engineering group can't be working around the clock, sweating details and encrustations, while the marketing folk have shifted their attention to the next product, assuming that the current one is temporarily out of their shop. If the salespeople are waiting to see the finished product before they start pushing it or are waiting for all the specs to be polished up for them, they'll likely ruin the entire project.

Everyone — every professional, every technician, every line worker — has to understand what makes the product cycle go, and what hurts it. As a manager, your job is to tell your people what helps, when to stop beating dead horses, or even when it's okay to wait for results from elsewhere. On the low-tech assembly line, it's easier to keep your nose on your own grindstone, the nature of the product cycle is better known, the technical risks and uncertainties are fewer. ♦ In your company, the connection between your job and good product is not nearly as clear.

One product I worked on, a big piece of chip fabrication equipment, represented a large investment in science and engineering. While the mechanical, optical, and electromechanical elements were complex — in fact, overly complex for the machine's function — the software was ridiculous. Anybody could see the hardware's state of development, but the software was handled like the emperor's new clothes. The problem appeared to be that the manager in charge did not understand the size or complexity of the job, or its relationship to the development cycle for the whole machine. Instead of begging to reduce the number of bells and whistles, he kept adding them. The tool could have been run by the equivalent of a couple of dozen clock timers but became mired in a hundred thousand lines of questionable code. My own feeling is that the perpetually unavailable software destroyed the product and subsequently the entire organization.

## COMPETITION AND ACCELERATION

You're halfway through a new product cycle, and Quicktek, your closest competitor, hits the streets with what amounts to the same product. Now what? Any technical person will have one of two reactions: "Let's finish our product because it'll be better" or "Let's drop this and beat them on the next generation." The decision to do either of these things probably isn't yours to make, as it is an upper management function, but it will be made, and you will have to alter your plans to suit. If you continue with what you're building, you'll have to speed up, or possibly hang on some more features. If the product is dropped, you start over. Competition's foot is on your accelerator, and its hand is on your company's wheel!

If you'd like to avoid getting whiplash from this kind of thing, you have to work harder on keeping informed about competitive developments, so that you can make these transitions smoother and more efficient. This information is detail, low-level detail that you and your group gather from conferences, literature, gossip, and other less-approved sources, such as spying. Your boss isn't going to get it, nor is his boss. Gather and pass along this stuff (sometimes called *competitive technological information*, or *CTI*), and there'll be fewer wrenching surprises.

## PROOF OF CONCEPT

This step in product development can be glossed over easily by wishful thinking or schedule acceleration. It's easy to go directly from an idea to cutting metal, especially if the idea has a powerful backer. If you get put in the position of doing product engineering work on an undemonstrated, unprototyped concept, you're in trouble. If all things were ideal, there would be no hand-offs without proof of concept, but it will take just as long to do the actual development as the prototype, so . . . .

Fortunately, you can speed up the proof of concept step in many ways. Simulation is increasingly possible for electronic and mechanical designs. Specialist organizations, such as the larger consultancies, have the intellectual and physical tools for rapid evaluation of complex concepts. Experimental design methodology is also emerging as a universally applicable way to get at the most relevant data in the least time.

Above all, cleverness is required. You have to be able to guess which aspects *need* to be tested and which ones don't. You have to know how rough a prototype can be and still represent the performance of the finished product. You have to know what's really new and what can be extrapolated from other work. You have to know enough about statistics to know *how many* prototypes are necessary.

All of these methods above will reduce the time needed to prove a concept. What increases the time? The worst time eater, in my experience, is allowing the most knowledgeable idea originators to hand off the proof of concept work to the less-informed, such as technicians and junior staff. The latter people often do not know what's important and what can be sketched in. Machine parts wind up over-toleranced or carefully finished, one-shot test equipment is made as if it will be in daily use, and so forth.

Another alligator is reinventing the wheel. The information you need may already exist in or out of your company. A day spent digging in databases or phoning around can easily save weeks of labor. Unfortunately, wheel reinventing goes along with NIH — "Not Invented Here" — which is a widespread concept that everyone else's data is suspect and must be tested. In some fields, all sorts of well-known, established phenomena get tested over and over again, often because of the lure of successful experiments!

## EVALUATING MANUFACTURABILITY

You have the working prototype in hand. The next task is to see whether it can be made, in your plant, as a product. In your business, this may be a trivial activity. A plastics molder can look at a new design and know how the part will be molded, on what machines, using what materials, and how quickly. ♦ In high tech, manufacturability is not a given. When a fundamentally new species of product is under development, all sorts of serious questions have to be answered:

- Can it be made with high enough yield?
- Can it be made with present equipment and technology?
- Can it be made cheaply enough?
- Should it be made somewhere else?
- Should the design be changed?
- Who has made a similar product or component?

The high-tech business pundits have been complaining lately that manufacturing, as a career, has not had enough emphasis and that it is not attractive to better and more creative people. This is quite true. The schools either do not cover this curriculum, or if they do, they wind up being too theoretical about it. The very creative job of evaluating the manufacturability of a new product may not be done well or efficiently. The result is unrepairable products, too many components, quality problems, overcritical adjustments, and similar sins. So what can *you* do about it?

♦ One good approach is to avoid total hand-offs. Following a product through its development is a way to stay focused — on the concept, on the marketplace need, and on the design rationale. If you or one of your group can go with the design into the manufacturing phases, a great deal of time can be saved. Documentation never adequately covers the thinking that went into design choices, so it's a pure waste to have someone else try to guess. Texas Instruments successfully institutionalized this approach, managing to avoid the reluctance of ivory tower workers to get down to the factory floor.

## BOTTLENECKS

Bottlenecks in the process are everywhere. Identifying them is part of the normal planning process. A bottleneck stands out whenever you can't get a commitment from another group, a supplier, or a service. Reducing or eliminating high impedances is a different activity . . . and very risky. GTE, for example, had a research department that was always crippled by its own shops. A six-month turnaround time on a simple part — a plate with a few holes in it, say — was normal. Nothing could be built in a reasonable time, and company rules prohibited going to outside vendors. This morass was a result of poor management. I've seen the identical situation develop at enough other places, and I have a theory about how it happens.

Most companies have, over the last 20 years, reduced or eliminated overhead facilities like shops, storerooms, and equipment inventories. Keeping machinists, glassblowers, electronics builders, and other prototype and model people around simply became too expensive. In addition, people learned that the Japanese avoided underutilized resources like the plague and had developed and proved out their methods. Unfortunately, as companies slimmed down, they did so with no forethought, and the shops fought back with self-created bottlenecks: long lead times and phony rejection (on quality grounds!) of outside-sourced parts. The purchasing departments created excessive paperwork in order to protect their jobs and established policies that attached purchase requests to mandatory rummaging through warehouses filled with unusable junk. For a project engineer, escaping these bottlenecks could have been a matter of obtaining high-level permission to go elsewhere for the same shop services. However, most of the engineers were afraid to make a fuss and learned to tolerate the slow-motion environment.

Bottlenecks created by company culture are largely intractable. They exist wherever there's a benefit to foot-dragging, wherever someone is looking to make political trades, and wherever someone thinks that they're saving their own job. Bottlenecks persist because they do not directly annoy upper management, who also do not yet have the skills to rectify the problems.

Minor or temporary bottlenecks aren't as bad. If you spot a way to save resources or time by changing an administrative procedure, a communication delay, or even the physical arrangement of equipment, suggest the solution to your boss and see what happens.

## DEAD ENDS

When you're trying to get something done on an ultratight schedule, every fruitless dead-end effort is added frustration. Unfortunately, high tech is research and development, and dead-end experiments are the rule, not the exception. You can't avoid spending a good fraction of the available time and resources on failed approaches.

♦ What you can do is know when to quit and abandon an effort and, more important, how to turn your staff around for a new direction. There's a lot of inertia in the staff, in equipment, in administration. A skilled manager will be able to anticipate redirection, get the organization prepared for it, and make a fast and smooth transition. Specific techniques will be covered in chapters 6–8. The next few paragraphs present some guidelines.

Don't let any concept become sacred. If you promote the current work as the only and best solution, you create a mindset that resists abandoning the effort. This is one of the easiest ways I know to burn out your staff. Everyone should be interested in and committed to the work, but nobody should become so personally tied to it that dropping the concept becomes a problem.

The decision to shut down an effort comes from you, consensus of your people, and upper management. The reasons may be clear and compelling, or they may be arbitrary or even bogus. It doesn't matter. Your job is to go with the decision and not look back. Allowing someone to continue on abandoned work after hours or part time "just in case" may strike gold, but more often it's bad management because you're instilling doubt everywhere. This doubt causes everyone to be just a little slower in making changes. It's not easy to kill anything, and you may be tempted to save an officially abandoned project, but you have to consider the effects on your group.

Professionals, especially creative ones, become personally identified with their work. They publish it, become known by it, and form their reputations on its quality. The closer to basic research you are, the more important this is. Thus, when you make the decision to drop something, you will encounter tremendous, often irrational, resistance. Sometimes people will even quit over a redirection. A good manager, believe it or not, can, by careful preparation and attention to the staff, bring anyone through this wrenching experience. Think of the process as similar to a divorce. Do it cold, and the aftereffects will linger forever. You're the therapist.

You also have to be ready to install new equipment, handle administrative changes, and revise your plan. If you prepare for it, you won't waste time. One of the worst situations is to terminate an effort and have everyone sitting on their hands for weeks while the new direction gets organized, obtains facilities, or waits for funding. This is wasteful and severely demoralizing. Your group isn't a machine that can be shut down or warehoused; it has to operate all the time in a purposeful way.

When you look at your plan, each milestone represents a place where you may have to make a decision to end one path and take a different one. A current, detailed plan helps you to prepare for, and get through, the problems connected with redirection and dead ends. This is why it's a good idea to make the plan a relatively public document. When everyone knows that a decision point is coming up, there's much less shock when a course change results.

## MAKING YOUR OUTFIT MORE RESPONSIVE

Responding to new direction varies from group to group, division to division, and company to company. Sometimes it's fast and easy; sometimes it can't be done without scrapping the organization and firing everyone. The trend in high tech, unfortunately, favors the latter: junk the organization or even the whole company, and start something new from the debris. This is wasteful and, in the long term, destructive.

You can do your part to buck the trend by having a group that is fast, flexible, and able to become productive quickly. Half the secret is in having smart, adaptable staff. The other half is in the way you manage them.

A group consisting of narrow specialists who have been hired solely on the basis of experience with a particular slice of the technology pie generally is a bad bet when change is necessary. You're better off with people who have demonstrated flexibility, know how to do research, and are motivated by success. When you announce that you're giving up on Hall-effect devices and concentrating on laser diodes, instead of blank stares and apprehension from your magnetics expert, it would be nice to see a generalist automatically start reading up in the new area.

The management aspect of quick change is just as important. You have to prepare for change, explain it, and motivate everyone to embrace it. How it's done depends on what personal style works best for you. An autocrat will just issue orders. The entire motivational exercise will consist of "Do it or get fired!" A better manager will try hard to get a consensus established on the new direction — its utility and validity. The motivation also depends on what seems to work. Personal opportunity to work on something new is attractive to nearly all of us. Others look to company success, profit sharing, raises, and bonuses.
♦ Each of your workers responds to different motivation, and you have to sense what it is and provide it.

Equipment and facility resources are impediments to change. These days it is not sound business practice to keep a stock of general-purpose equipment around when it's not in use, and it's also no good to have shops. Both of these practical realities make change more difficult than it once was. "Quick and dirty" now becomes "slow and premeditated." The situation forces you to learn new ways of getting things done, such as scrounging, bartering, and using external facilities at universities or even competitors. The one thing you *can't* do is to wait very long for these resources. You, your staff, and your management must all be aligned on the necessity of finding unconventional ways to get work done.

Small companies are considered to be more agile than large, bureaucratic ones. There is less administrative burden, fewer people have to sign off on things, and the company culture, having no history, is oriented toward the future. Small companies should have distinct advantages when rapid change is required.

However, large companies have more available cash, more internal resources, and much more leverage with partners and suppliers. When Big Blue wants anything, suppliers will burn rubber to get there first. Also, when sheer intellectual power has to be brought to bear, the big companies either have it or can attract it.

Obviously, there are advantages and disadvantages in every company. The trick is to be able to understand the organizational structure well enough to exploit the advantages you have and skirt the disadvantages.

## HOW TO SPEED UP — A SUMMARY

- Understand product cycle time.
- Keep everyone informed on the "big picture."
- Know your competitor — use this knowledge.
- Know what's manufacturable and what isn't.
- Find bottlenecks and eliminate or bypass them.
- Understand dead ends and how to change direction.
- Find alternative ways of responding faster.

# CHAPTER 6

# How to Create an Exceptional Team

This chapter is all about how people interact at work. It contains the kind of material a scientist or engineer may not value enough to study. Other managers may turn to it first. Engineers and technologists historically have had a low opinion of soft, human issues such as social dynamics and psychology. The common perception is that, if not entirely nonsense, interpersonal interactions are at least mysterious. We'd like to believe that work gets done without fuzzy, erratic, human dynamics, and that nobody needs extra motivation to do a good job. This bias is what makes some technologists such poor managers . . . until they learn better.

At its core, the management job is controlling and guiding other people. Motivation, delegation, communication, facilitation — none of these crop up in an engineering education or show on a spec sheet.

Because you're in high tech, not only must you learn some soft science, but you've also got some very unusual people around who respond to obscure motivations. They can be asocial or antisocial, or they can have lopsided personalities. Not only do you have a "people job," but it is a specialized one.

## THE TRIBE

You and your group form a tribe. You're the leader. As primitive and simplistic as this concept sounds, it seems to be valid. Your organization can be as important to its members as family and friends, and frequently it becomes more important. The pressure and stress of a high-performance workplace can easily disrupt marriages and other relationships.

Professionals get most of their self-esteem and sense of worth on the job. It comes from the work they do and the success of the company. Job mobility has scattered most of us around the country, far from family, and has destroyed our roots. The people we work with become our entire social environment. When we work twelve- to sixteen-hour days, spend weekends fighting fires, and have the high level of commitment that is necessary in our kind of business, we may not even *see* other people, let alone develop any relationships. In fact, one of the hallmarks of the heads-up enterprise is an astounding divorce rate. I don't endorse this arrangement, but it's there and needs to be understood.

So you have a tribe. When it works, it's a synergistic, cooperative group whose total output is far more than the sum of the individual contributions. When it doesn't work because of poor management, the group becomes an internally competitive, destructive, unresponsive waste.

Your only justification for being a manager, remember, is leverage. Your actions increase the amount and quality of work performed by others, to a degree that pays your way. When the leverage isn't there, you might as well not even be there. In some tribes — for example, basic research groups consisting of almost independent, self-directed people — the leader's leverage is minimal, and he or she is able to look like an individual contributor.

---

**How to Make the Tribe *Your* Tribe**

- Show that you care. Help someone out with a nonwork problem, such as getting settled, finding goods and services, carpooling. When serious problems, such as drugs or family strife, are hurting your people, be willing to show support.

- Spend nonwork time with your people. This could be going to a Friday group happy hour, playing weekend tennis, bicycling, or attending sports or entertainment events.

- Never close your office door, if you have one. For that matter, encourage people to call you at home if a problem crops up. If an engineer is working late, management is working, too.

- Practice a little "management by walking around." You should show your face at each desk once a day, when possible. When you do this, be careful about not trying to do an employee's job or dispensing too much wisdom. You're there to learn and support.

- If someone gets into a tussle at work with another group, support your staffer even if it means catching some flak yourself. Of course, never back a seriously wrong position.

---

## WHAT IS A GROUP?

Your group is defined by its function in the organization. Somebody has given it a label, such as "manufacturing engineering" or "advanced research." It has a distinct and specific role in the company. Inside the group, everyone also has a distinct role, which isn't the same as the job description. These roles relate to the internal structure of the group, which mostly organizes itself, with a little guidance from you.

This internal structure, if you try to represent it on a piece of paper, consists of *roles* and *interactions*. A circle can represent each person, and dotted lines can be used to show who works with whom most frequently. Other lines or symbols represent attractions and repulsions. If you draw a line around the whole group, you show its *boundary*. This boundary might include people who don't work for you, and it might exclude those who do but who are disconnected and externally focused. These are not functionally part of the group.

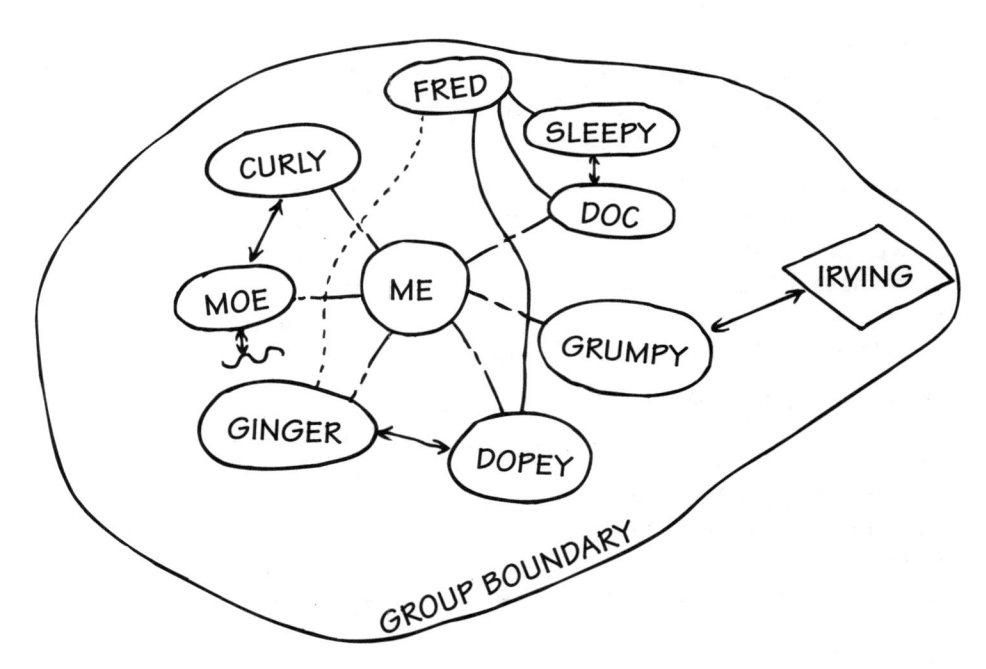

Your company consists of many such groups, or tribes, each with its own identity. They communicate with each other and with the outside world. They trade services, information, and resources all the time. Some of this communication is under your control, some isn't.

Your group is also like any other living entity in that it tries to preserve and perpetuate itself — if it's healthy. This self-preservation fundamentally consists of trying to do a good job or make a good product. It also involves competing with other groups for attention or resources. What's important to your understanding of group dynamics is this:

*A group will fight like hell to keep itself the way it is.*

### How to Get Changes Accepted by the Group

- Each change has to be preceded by valid motivation, discussion, or consensus. Good people always want to know why.

- Whenever you can crow about a previous change that worked out, do so. Try to establish a reputation for success in small things. If the group thinks that you're making changes just for show, your goose is cooked. Brand-new managers, for example, can't resist making frivolous changes, such as moving furniture.

- Share credit for successes. Don't hog it all. Don't be like the publication-counting academics who feel obliged to hang their names on every piece of work that comes out of their shops. Make sure that praise for a successful change is properly distributed. This will encourage people to take the risk of making suggestions.

- Encourage stability and discourage rituals. The last thing you want to see is someone needlessly repeating an approach because "it's the way we do it." Make it clear, however, that no matter how the work changes, the group will remain together. Five years from now, your group could be intact and functioning, doing nearly anything. The details of last year's work will be forgotten.

- The bigger the change, the bigger the payoff. When a change is going to be obviously disruptive, you have to help the group look beyond the change to a clear reward. Don't lie. For example, if you're dropping the entire project, don't minimize the effect of having everyone abandon good work. Instead, explain how the next project will personally benefit each worker.

## A GROUP ORGANIZES ITSELF

If you have inherited a group, it's already organized. Each member has some idea, possibly outrageous, of his or her role. All the friendships and hostilities are firmly in place. In fact, within a week or so of formation, the group organizes itself and then becomes resistant to change.

You have a limited time in which to establish yourself as a leader when you become one. Otherwise, the group will form other ideas about your role. In the worst case, they can decide that you're some nominal administrator, a leader only on paper.

When you bring a new hire into the group, the new person also is evaluated quickly and is given a role to play. This role could be something unexpected by you or the new hire. Sometimes a new hire suddenly quits after a few weeks or a month. If you don't know why, it's usually because the group either spit the newcomer out or put him or her into a less important role than expected. (In this situation, it's a good idea to try to find out exactly what happened, so that it doesn't happen again.) With each role, there are *expectations*. Everyone believes that they will get certain treatment, rewards, recognition from the group, from you, and from the company. Like roles, these expectations are always off-base and are usually wishful thinking. Part of your job in preserving the health of your people consists of trying to keep their expectations close to reality. You don't want the difference to cause frustration or burn-out.

How much you can do to influence the organization of your group depends on skill, experience, and the particular mix of characters you have. I expect creative, highly-educated technical people to be especially strong-willed. They can be difficult to control or influence. I expect challenges on decisions, severe hostilities, destructive competition. I also expect that some will be so remote that they have no sense of the group social organization and could care less.

As soon as you can, try to figure out what organization is there, what all the roles are, and where you fit.

## ANOTHER ANALOGY

The punk analogy of chapter 1 continues: Think of your group as a molecule. (A physicist, of course, will invariably ask you to consider that anything can be understood as a sphere! Chemists differ in that they like clumps of spheres.) A molecule consisting of immutable individuals (atoms) connected by forces and interactions (bonds).

When the atoms first come together, they sort themselves out according to these forces and form bonds of various strengths. Once this has happened, the molecule is fairly *stable*. It requires exactly the right *excitation* or *chemistry* to change the structure. Only a few new arrangements are possible. Even for simple molecules, the number of interactions is very large.

Your four, five, or ten people are also individually immutable and will form bonds with others only in certain ways. The group itself will be stable and may or may not accommodate additions, reductions, and rearrangement. When groups condense into companies, they fit only in certain places. Forcing them into unlikely structures causes instability.

♦ In some management theories, or cultures, human beings are considered as largely interchangeable. A mechanical engineer is a mechanical engineer. The comic book notion of the Japanese factory is an example. In our culture, we like to think of ourselves as irreplaceable individuals, not anonymous drones. The truth probably lies somewhere between these extremes. Maybe we're interchangeable atoms of a hundred types or maybe we're all really individuals. Either way, you've got some work to do to figure out the composition of your group and what it means.

### How to Learn about Roles and Expectations

A role is what the group expects someone to do. The clown is expected to provide buffoonery, and the creative one is expected to create. *Nobody knows what his or her own role is, so you have to ask someone else.* The best way is informally, as in, "Nancy, I'm working on a proposal. Who do you think would like the (process, design, modeling) part?"

Expectations are group concepts and individual concepts. The group ones come out when you investigate roles. The individual ones can be gleaned from performance reviews, informal conversations, and history. History is when you learn that Moe has left his last four projects very soon after the initial, creative stages were completed. Moe expects his work to be constantly on new subjects. A good manager has to avoid crushing anyone's fragile expectations, even while she is suggesting new, and more supportable ones. Moe can be helped to expect and want the rewards of successful, if routine, development.

Some of us expect failure. Pessimists may be right, but they cause damage. They're hard to spot because most rational people know better than to be negative in public. There are quite a few pessimists around, however, and you want to keep them away from tasks they can't embrace. Asking the group to estimate probabilities of success on a project is a good way to find the nay-sayers. Once you have identified them, it's up to you to give them enough reasons to abandon their dour outlook.

Pay attention in meetings. Try to listen more than talk. Identify the roles people play. It's a safe assumption that there's counterpoint: leader-follower, outspoken-timid, optimist-pessimist. The ones who don't talk are as significant as the ones who talk all the time. The ones who don't contribute in meetings need encouragement. The ones who are obstructive need to be reoriented.

## FACTIONS

♦ Those who work well together, or merely like one another, form little *factions*, subgroups of the main group. These subgroups don't show up on your organizational chart, but they are extremely important to the functioning of the group. These factions are competitors for attention and resources. One reason a manager is given so few (usually fewer than ten) professional employees to manage, is that above this number, factions will be certain. Instead of one manager and a group of subordinates, you have several self-contained groups, *each of which manages itself*. The net effect is not good. Our people will even form factions of one, setting themselves outside the group.

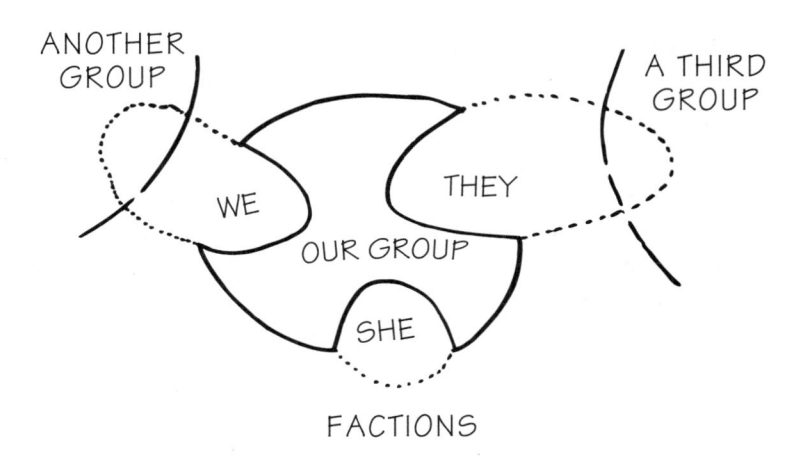

FACTIONS

Beyond your own group, different parts of your company also can split into factions. Product engineering and manufacturing engineering frequently see themselves as competing over the same turf and for the same resources. One research group may ignore or attack another one. Even very large organizations, such as divisions, can turn into squabbling factions.

Managing a faction is just the same as managing a group. It gives you a harder job because you have to satisfy more than one bundle of expectations. Because they compete, this means that one faction will always feel slighted. The other problem with factions is that they often form bonds far outside the group. The productive, publishing scientist whose career demands a high level of interaction with a peer group spread all over the world is a faction who owes you very little attention.

### How to Reduce the Harm of Factions

Factions are harmful when they compete with each other so much that resources are diluted and work is duplicated. Also, it's not good having someone spend time trying to prove someone else wrong. So, the first rule is: one task, one person. Having two people do the same job is like giving two of your kids the same name.

Stir and blend your factions. Interchange their members on projects. For example, if a professional always works with the same technician, the two of them become a faction. If this appears to be a problem, swap the technician. If it is a productive arrangement, leave it alone.

Factions that are oriented outside the group may be acceptable, but they reduce the available resources you can apply to a problem. You can pull them back in, if you become good at motivation. If you have a worker whose job is unrelated to the main thrust of the group, as often happens in research, it's a good idea to look into transferring him or her out of your group.

♦♦  Establishing worthwhile competition is another method of handling factions. An intractable problem can be tossed out between two factions, with a reward for the best solution. This is as effective as it is dangerous. Effective, because competition makes people work around the clock. Dangerous, because it divides the group further, possibly diluting resources and duplicating work. Once in a while, though, it's useful.

## THE BOUNDARY

In most companies, other managers can draw upon your group's services or borrow the staff. When a company organization allows managers to operate across group boundaries, we call it *matrix management*. This is a popular system, and it works well in small, fast-moving outfits. However, having overlapping bosses presents some managerial problems.

The first alligator is that other managers can contradict your decisions and otherwise step on your toes. If you're just learning the ropes, you can easily lose the ability to direct your own staff. That is, you can lose your authority, while remaining responsible for the staff's output.

The second boundary-crossing problem is that you have much less control over communication in and out of the group. This means that you know less about the information channel than you need to.

The third alligator is that you can lose staff to other managers. Another group may become attractive to your people if it has more resources, a more congenial atmosphere, or a more powerful — hence faster-rising — boss. This doesn't happen as easily in nonmatrix organizations.

How do you figure out where the boundary is?

The only way is to have a complete accounting of who communicates with whom and how much. If you have a mostly electronic office system, this information might be available. Then again, a number of us try to make sure that it isn't. You may have resort to just paying attention.

---

**How to Make, and Keep, a Good Boundary**

- Make a physical boundary. Very few of us like inhabiting endless, anonymous, modular-furniture rabbit warrens. Having a distinctive piece of physical turf (it may be as minimal as a building corner or a field office) can lend a sense of identity to a group. If you have the opportunity, arrange your space so that everyone is in the same general area and can see each other.

- Reinforce the social boundary with group picnics, celebrations for delivering objectives, and so forth. One of my employers, Balzers, Inc., broke out a case of champagne when a major system shipped. Solidarity with bubbles.

- The secretary used to be the hub of a group. As secretaries have been replaced by chunks of electronics, groups lose a certain focus and identity. The only replacement for this focus is you. This is another reason you need to be visible and accessible. You don't necessarily have to put your desk next to the coffee pot or the copier, but it wouldn't hurt. One manager I knew at Perkin-Elmer insisted that his desk be in the center of his turf.

- Make sure that other managers think of your group in terms of its function, if that's the way your company is organized. Try to use the word *we* when referring to the group, not *I*.

---

## EXAMINE THY NAVEL

Authority and control are touchy subjects, especially when some of the people working for you are smarter and more experienced than you. Still, you must motivate others to do actual work and hold them responsible for it. When you're doing this, it always pays to ask yourself: "Is this the best approach, or am I just displaying authority?"

If you're not sure about the answer to this question, ask someone else for an opinion. I'd suggest asking your boss, because his perspective may be better, and because he will feel flattered that you asked. In a group that runs on consensus, you could even ask everyone.

What do you do if you find yourself making mistakes because you feel compelled to issue orders? You could make a conscious effort to delegate more decisions to your group. You could also profitably read more about management techniques. The one certainty is that your intelligent and skilled staff will not respond as well to orders as to motivation.

## MOTIVATION

Thousands of books exist on this one subject; thousands of academics, consultants, and lecturers earn a living teaching the elements to business and government. Where it is overwhelmingly important — for example, in the Army — motivation is studied and refined continuously. The basic reason for this attention is the understanding that a well-motivated worker produces more and better work than one who isn't.

There is no question that the concept is valid. However, the myriad of theories on how to motivate can't all be simultaneously applicable. The main point of this book, in fact, is that high-tech management is fundamentally different than other kinds of management. I think that you can learn something from *any* kind of theory, however, and I recommend that you read a little about Maslow, McGregor, and Theories X, Y, and Z, which you will find amply described elsewhere.

Technical people have unusual motivations. Sometimes these are so remote as to be incomprehensible. How about your buddy who spends every waking moment peering into a computer screen, living off anything that comes out of a vending machine, preferably something heavy with sugar and caffeine. He isn't doing this for money, approval by peers, promotion, or any of the normally cited reasons. What's more, this apparent workaholic isn't necessarily doing your company any good.

The scientist who is wobbly from fatigue in the lab at four in the morning, presumably while in search of another data point, could well be engaged in self-flogging because of an uncomfortable situation at home. Maybe he doesn't want to be there. This also may not be doing your group any good.

The main message here is that, in high tech, you are managing an extraordinarily diverse collection of human beings who march to many drummers. It takes understanding and effort to make *your* drumbeat heard above the rest.

*Stein's Theory of Technologist Motivation is very simple: Tell everyone what this week's motivation is, and why it's important to them, to you, to the project, and to mankind in general.*

Will these recalcitrant, difficult people respond to this? Maybe, if you're believable as a leader. It's a better idea than trying to grapple with all the group interactions and finding all the individual motivators.

◆ The key to letting your people motivate themselves is your credibility. In order to be believed and trusted, you have to make your own motivations visible, and they have to be palatable. We don't trust the boss who is clearly interested only in her own success. We also tend not to believe people who are motivated by concepts we can't understand. Before you can tell someone else what this week's group motivation is, you first have to explain your own.

Here's a list of motivations (in no particular order):

- Money for food and shelter
- Approval by coworkers
- Satisfying work
- Participation in company success
- Opportunity for fame
- Approval by the outside world
- Nice toys to play with

You can try any motivation that seems to work, but bear in mind:
*If you promise what you can't deliver, you'll lose, eventually.*

---

### How to Motivate People You Can't Understand

◆ • Everyone comes already motivated. You can substitute, alter, replace, or modify motivations, but the one that works the best is the one that is already in place. The trick is to find out what it is.

◆ • Utilize the substantial power of the group. This is called the *group norm*. Once you get the majority headed in a particular direction, it might catch on. This works well when you are not experienced enough to deal with individuals.

• Emphasize the temporary nature of the duller work. By looking forward to the next project — which, of course, will be challenging and rewarding for all — you help the crew invent their own rosy pictures of the future. As bizarre as some of these images may be, they all serve to get the current work completed.

- Talk to people about projects they've enjoyed in the past. See if you can understand why, and see if there is any correspondence to the project at hand.

- If there's someone you find particularly unfathomable, ask him for advice as much as you can. Sooner or later, he'll tell you what he wants.

## DELEGATION

There are two kinds of managers: those who can't delegate and good managers. Look at it this way: You pay your way only by leverage. More work is done by benefit of your presence than would occur otherwise. If you have five people, this increase in productivity has to be more than 20 percent, for example. Each task that can be delegated and isn't costs your company more.

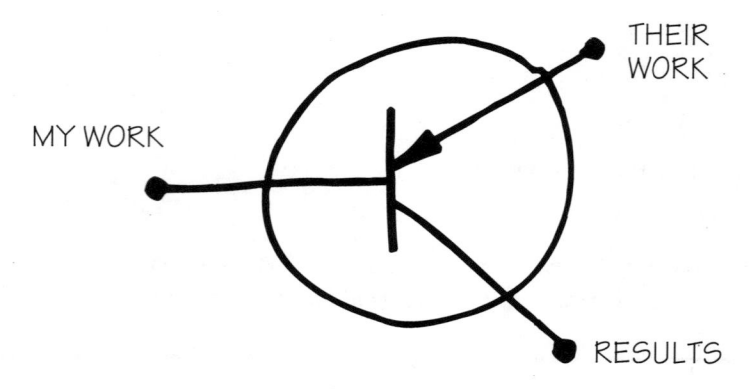

LEVERAGE OR DELEGATION

The biggest problem with ex-bench people promoted into management is keeping them from trying to do their old technical jobs instead of delegating the work. Some of the time we fall into this trap because we are unsure about management but know that we are good at technology. Some of the time it's because the manager's promotion came from being the best technical person, promotion being the only available reward. Often, the manager's "people skills" are so bad that effective delegation is impossible.

Even though you're a manager, you cannot delegate all the technical work. The rule of thumb is that the higher you go in the organization, the more you delegate, so that the president, for example, delegates 95 percent of the work that reaches that level. You may spend as much as 75 percent of your time on technical work, but that represents only a small fraction of the total work that arrives at your desk. Therefore, you delegate the rest, and you have very little time to administer it. You also may well be the most technically competent person in the group, which hampers your ability to delegate.

### The Lure of the Bench

Harry thinks that the design on the breadboard isn't quite right and that he's not hearing accurate reporting in meetings with his group. So he starts spending several hours each day at the bench, with his own hands on the hardware, doing what he knows best. Soon, several hours turns into all day; the administrative stuff is ignored. The rest of the group is wandering off course, and Harry's boss can't find him. Harry will be surprised when his group is reorganized — under someone else.

What should Harry be doing? First, he has to determine whether or not his people can do the job. It's not important that Harry has the skills. If nobody can handle the design, Harry needs to hire a new engineer. If the skills are there but the reporting is wrong, he can address that problem by itself. If help outside the group is available, Harry should arrange for it.

### Things That Can Be Delegated

You can delegate two things:

- Responsibility
- Authority

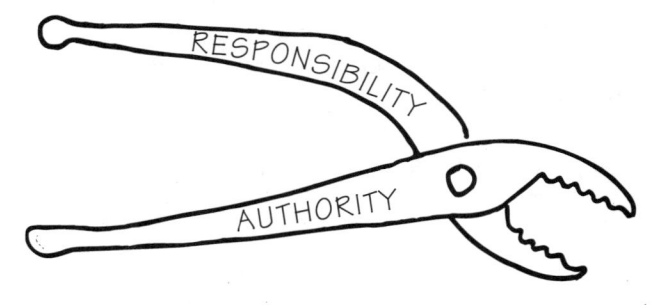

When you give anyone a task to complete, you're delegating responsibility. The big chunk of responsibility is ultimately yours; you are handing off just one portion. Your project plan, if it breaks down a program into assigned tasks, is a chart of delegation of responsibility. The names put against particular tasks are people who agree to deliver the objectives. This is a contract between you and your employee. It includes, one way or another, a promise to deliver something, or else. In turn, if you've picked the wrong person for the job, and that person can't deliver, you're the one who will be responsible to your boss.

Here are a few suggestions for delegating responsibility:

- Never give the same task to more than one person.
- Make sure the right person gets each task.
- Make sure that your reasons are clear.
- Delegation can be a form of punishment: Make it clear.
- Don't delegate competing tasks.
- If it's a reward, make sure it's identified as such.

These suggestions can be implemented most easily by remembering that delegated responsibility is a contract and that your subordinates are your subcontractors. You would never give the same purchase order to two suppliers, with no indication as to who is responsible for delivery. You also would like to look at past performance in a highly specific area before buying anything from a vendor. You also try hard to be as specific as possible in describing what you need. You can delegate to your group in the same way, so that responsibilities are appropriate to abilities, unintended competition isn't created, and the deliverables are as specific as possible.

In addition, you should consider two interesting social aspects. Responsibilities are often used to punish or reward staff. The nonperforming worker is given an unattractive responsibility, perhaps involving record-keeping or other janitorial tasks, or the technically bland part of a development job. Your best group member may be given the leading-edge glamour job, representation of your project in public or at corporate reviews, or contact with the most important customer. Whenever you intend an assignment to be a punishment or a reward, there will be no effect unless your intent is communicated. *Tell* your star performer that "because you are so important to this effort, you are being given new, and vital, responsibilities."

When you delegate authority, you're assigning *licenses to act*. This may be signature authority on purchase orders, unsupervised travel, planning, and other resource allocation.

Every new manager is liberal in handing out responsibilities but tries to keep all the authorities. Responsibilities are work; authorities are power. Try hard to avoid this, because, in effect, you're asking your people to build a wall but not giving them access to tools.

Here are a few suggestions for delegating authority:

- Make sure that it's yours to delegate!
- Try to keep authorities and responsibilities parallel.

- If authority is very important, your company is crumbling.
- Do it in writing.
- Make sure other managers agree with what you've done.
- Make sure it doesn't violate company rules.

♦ A new manager is usually aware of what the group's responsibilities are but is shaky on the subject of the associated authorities. Can you send someone to a customer site on your own? Can you allow your people to do the same without asking you? What authorities can you delegate without violating company rules or stepping on your boss's toes? Do you have to get other managers at your level to consent?

♦ Once you know what you are able to delegate, you have to pair responsibilities to authorities. It's tasks and tools. It's also important to everyone's sense of worth to the organization. The vice president of manufacturing who needs corporate permission to rent a car for a day doesn't feel as important as his title and may even become less decisive or adventurous than the job requires.

♦ Excessive concern for strict levels of authority is endemic in larger companies and in ones where business is poor. The dividing line between appropriate and excessive is crossed when you can see unreasonable time delays for signatures, names added to sign-off lists, and similar unproductive practices. Large organizations always develop bureaucratic encrustations that account for inefficiencies. As always, you have to help solve the problem or become part of it. In other companies, the same symptoms of authority hoarding usually mean that money is becoming a problem. The cure for a lack of money is rarely to be found in counting it more carefully, but companies on the skids seem to do it anyway.

Finally, when you tell Norma that she can buy mainframe time without having to ask you, do her a favor and write a memo to the computer center manager, with a copy for Norma. An informally delegated authority, combined with anything but the most confident outlook, is a recipe for stress.

## PORTER'S COMPLAINT

Porter heads a group of scientists at a national laboratory. Aside from the other problems that accrue from working for bureaucrats, Porter has trouble with authority: He doesn't have any. All he has is responsibility. Every dinky purchase, every external communication, every bit of travel, has to be approved by separate, faceless, and remote organizations. The result is that planning is impossible. There is no way to guarantee that needed resources are available at the proper time. At first, the problem looked budgetary, because his group had to operate with excellent salaries but no other money. Now, Porter thinks that it's purely administrative. Porter is looking over the fence at the private sector but is afraid of losing the security of the government job.

Thousands of us are in this particular bind, and not all in national laboratories. Poor managers anywhere can distribute responsibilities while hogging all the authorities. Worse, they can assign tasks and demand results when nobody has the actual authority to make it happen. Then you see work that is mostly "pretend." The latest examples of the authority-responsibility mismatch can be found in consortia — big talk, good people, no authority over members, insufficient budget. The best I can suggest, which isn't very attractive, is that if you're caught the way Porter is, that you seriously consider whether making a contribution during your lifetime is important to you, and act accordingly.

## DELEGATION AND THE TIGER TEAM

The Tiger Team operates outside of all the normal chains of authority while having access to considerable resources. Because all the rules are broken, the members of the team enjoy freedom from administrative hassles — but as mentioned before, planning is especially difficult.

What the success of Tiger Teams shows, on the other hand, is that very good work can be done without most of the organizational structure that traditional managers think is so important, even sacred.

If you are on this kind of project, your job shifts away from delegation and toward facilitation. That is, your people are already focused and committed to an objective. Your job is mostly to keep them supplied with the tools and resources they need.

---

### ♦♦ How to Delegate Your Tigers and Live Dangerously

- Anyone who wants a particular task just takes it.
- Anyone who wants to change tasks midway can.
- If someone gets in too deep, find help.

---

This advice looks like a description of anarchy, or at least something best left to a highly experienced manager. The truth is that high-tech demands for performance and speed often put novice managers in charge of Tiger Teams or groups that act similarly. Your whole company might even operate like a Skunkworks, in which case, enjoy it while it lasts.

The problems that accompany lack of detailed delegation are many. The first is that you are never sure that you are in control of the work, or that it will get done. Another is that because formal reporting is ignored, you have more trouble measuring progress. Yet another is that people will be diverted from the stated objectives and will go off on their own, possibly accomplishing much, but also impeding progress.

♦ When you sense that you, the group, or individuals have bitten off too big a job and that you're in danger, you have the responsibility to find help. The help can come from other parts of the company or from outside. Because a tiger team is not bounded by company culture and regulations, it's actually easier for you to tap resources, such as consultants, who are otherwise difficult to justify. If the work is going well and nobody wants to "waste" time on documentation, you can borrow or buy the people you need. Need some machine shop work? Go with the fastest source, and don't bother to route it through purchasing.

To keep the problems in check, keep your communication channels open, whether through meetings, personal contact, or informal reports. Management by walking around is most applicable in this situation. First you ask how the task is going; then you ask how you can help with resources.

## MAKING MEETINGS WORK

You can't get rid of meetings. They are your primary method of communicating with your group. Meetings also can waste unbelievable amounts of time. I have seen organizations so badly decayed that a lowest-level manager spends more than half the day, every day, in meetings.

The worst meetings are those in which the matter under discussion only affects one participant directly. Many expensive people are tied up, doing no work, so that one person can present data or get marching orders. You should always be alert to the possibility that some meetings can have fewer attendees.

Another wasteful aspect of meetings is the incredibly obtuse practice of taking, writing, and distributing notes (not including action items) to the people who were there! After all, if you heard something in person, participated in discussing it, and didn't need to take notes then, why would you ever need to see some abbreviated version a week later?

At one point in my career, the government research part, I actually enjoyed long, pointless meetings. They broke up the day, were relaxing, and offered the chance to give gratuitous criticism of other people's work. In the science game, business meetings like this are interspersed with lectures. Some weeks, no time at all is spent at the desk or in the lab. What changed my opinion was my notebook. I had been good at taking notes at every meeting I attended, and I had a log that represented thousands of hours of sitting. One day, I went back through all these notes to see how much of this information I actually used later on . . . the answer was depressing.

Fortunately, the lean and mean trend in high tech has provoked widespread interest in reducing meetings and making them more effective. There's a growing literature on the subject, because more of us are recognizing that the net profitability of most companies can be eliminated *with one or two weekly meetings!* The other factor finally putting a lid on pointless meetings is the acceptance of the electronic office and electronic mail. Although it's no substitute for necessary human contact, it helps separate the trivial data transfer from the social stuff.

Here are some methods for making meetings work for you:

- Each meeting needs to have a distinct, preannounced purpose. Periodic meetings with vague agendas, or just show and tell, are pointless — except in production areas, where frequent, even daily, status reporting is necessary.

- The show-and-tell meeting at the group level where everyone attends just to report on what they have done lately is expensive, even if it helps orient everyone. The bulk of the audience doesn't need to know these details or already does. For updating the manager on progress, it's lousy, too, because it inhibits the confessing of errors, doesn't allow the problems to be examined at their source (for example, the bench), and allows too much immaterial discussion. Once in a while, someone will toss in some useful and relevant information, but it's not worth tying up everyone else to do so. Don't have this kind of meeting.

- Brainstorming meetings are good every so often. In these, you put up a problem and let everybody take an off-the-cuff whack at it. Someone stands at the board and writes down ideas. Discussion is limited to the end of the meeting. That is, no immediate criticism is allowed. Brainstorming meetings are good to have when you are getting into a new area, have hit some severe snag that can't be solved by the delegated person, or when you want to make your group feel some commonality and it's too cold outside to have a picnic.

- In any meeting in which presentation of data is necessary, the presenter should be given a limited time — say five or ten minutes — to do the job, and *nobody* is allowed to interrupt. Questions come afterward. Very few of us are such good public speakers that we can give a coherent talk while being interrupted.

- Elaborate presentation materials should be avoided like the plague. We all have software that makes display transparencies, so there has been a proliferation of glossy materials. These don't add much. It's true that legible materials are nice, and graphs and charts are illustrative. But even with the automation, this is all work, and nobody is paying you to do it. Last month's graph, for example, will accommodate an inked-on update and still get the data across. Like memos, another avoidable disease, display materials cost much more time than most people think. It's not uncommon for someone to come in two hours early on the morning of a meeting to make a few graphs. This is not a good use of time.

- The whole group doesn't need to attend every meeting. It is possible to schedule meetings for just the needed personnel. You can do this easily in an electronic office or e-mail setting. The obstacle to customizing meetings used to be the difficulty of finding and getting messages to participants.

- Memos of meetings are supposed to serve two administrative purposes: They formalize what someone said and what tasks have been assigned. The overuse of memos is either the result of too much organizational fat or a means for inflating trivial achievements. Come on. We're certainly worth trusting on what we promised to do. Dump the big memos. However, action items are vital. An action item is a short-term assignment of a task that is not already on someone's plate. *Short-term* means up to a few weeks, or until the next meeting. An action item calls for new work, in addition to everyone's ample load. Therefore, they should be used sparingly and only when significant. Chasing frivolous action items probably isn't a budgeted category in your plan.

---

### How to Save Ten Weeks a Year

- Start your meeting on time, regardless of who's late — including higher-level personnel. If you don't, your boss will always show up late and will request a lengthy recap.

- End your meeting on time. Even better, end it sooner.

- Few meetings should be so long that everyone *has* to sit. Have standing meetings, preferably at the work site, not the conference room. In the semiconductor fabrication industry, for example, the Japanese were first to realize that chairs could be removed from clean rooms on the grounds of reducing particle count. Along with a few flecks and specks, they also removed lots of idle time. In addition, since gowning up and down is difficult, they learned to have meetings where the work was.

- When someone has to leave to take an important call or whatever, do not pause the meeting. Do not recap until after the meeting is over. Shift to a topic that does not concern the missing person, if possible. It does not warm the hearts of many supervisors to know that while they're out dealing with business, the meeting has gone into neutral.

- If there are hand-outs that duplicate what's being presented on a screen or board, you have divided everyone's attention. Present the visuals first. Give out the hand-outs later.

## COMMUNICATION

Meetings are one form of communication. To be effective, your group has to communicate internally and externally. When this degenerates into protective documentation, you have failed. Effective transfer of information is:

- Informal
- Constructive
- Sufficient but not excessive
- Facilitative

Informal communication is always freer and more accurate. At the other end of the spectrum is a Swedish doctoral defense, usually done in a courtroom, with lots of bystanders and your family watching the interrogation. Technologists are often too timid to speculate in a big meeting or in a formal report, but they can tell you exactly what's going on in the time it takes two of you to pour two cups of coffee.

Communications should always be intended constructively. A problem can be identified, but only so that it can be fixed. Gratuitous criticism, backstabbing, and pessimism don't help. A constructive communication can also be an advertisement for assistance with a problem. If, for example, you are unsure of how to arrange a ground plane on a high-frequency circuit, posting the question on an electronic board might communicate the problem to someone who can solve it.

Communications should be sufficient but not excessive. Like meetings, communications beyond a certain point can waste time. In addition, if written materials are overdone, you have discouraged personal follow-up, another form of communication.

If your group has a primary objective of synthesizing a plant growth chemical, it's not facilitative to write and circulate a long synopsis of work done elsewhere on insect toxins. Communications should facilitate the doing of the work.

The technical workplace continuously invents new forms of communication. In the space of a generation, we have turned words, numbers, and pictures into easily manipulated new media.

A design review that had a group of engineers marking up a hand-drafted schematic now can happen simultaneously at several locations, involve immediate simulations and changes, and can eliminate further reviews. We sit at our own desks and exchange data with colleagues and libraries. We pull information from increasingly vast databases and pass it around with astounding ease.

We also go to lunch with one another and talk.

In all forms of communication, there are both the loud and the shy. Just as there are some who will never spontaneously utter a peep in a meeting, there are those who will not go to lunch with the group. Minorities and women sometimes miss communication channels this way because they may feel uncomfortable or because they have been excluded. Because you, as boss, learn more by listening than by speaking, it pays for you to encourage everyone to become a frequent communicator.

High tech also is afflicted with jargonitis, which makes for bad communication. The writing we do is so heavily laced with the abbreviations and temporary terms we like that as little as a few months later, it rots into gibberish.

A manager likes to feel that she sees all the information generated or received by the group. This is less possible every day. We do well to define a few categories of communications that require our approval. These may be responses to important customers, materials that go to our bosses, or publications.

Your role in the group, as leader, depends on your being the focus for as much of the communication as possible. The hands-off manager who stays clear of the information channels becomes an outsider to the group. Once in a while, hopefully not too often, you have to tag along with your bosses and not your group. Make sure that you bring information back so that your group doesn't feel left out.

*Stay involved. Read as much as you can. Direct as much as you can to your desk. Eat lunch with your people. Go out with them.*

## ♦ COMMUNICABLE DISEASE: CASE STUDY

Anna's group meets every Thursday morning at 8:30 to discuss work in progress, plans, and administrative details. The meetings are reasonably short, well-managed, and rarely digress too far. Anna realizes, however, that something is wrong. It seems that three of the five people in the group routinely write up and distribute memos recapping and commenting on each meeting. This is in addition to the official meeting notes that Anna writes herself. She estimates that the one-hour meeting now costs about 18 hours of work, rather than the 8 that it should, not counting anybody reading all these memos. When she brings up the subject, the memo writers all insist that their "slant" is important and that they'll write after-hours to minimize the cost. What's actually going on?

The excess memos are a symptom of fear, anger, or strife in the group. The staff is sniping, covering their butts, and acting like squeaking wheels. They are doing this only because Anna has not succeeded in managing them into a cooperative, united tribe. Individuals no longer see the group's objectives as parallel to their own, and they no longer expect rewards for their work. Therefore, they feel that they need to compete for Anna's attention and denigrate each other.

Anna, although not experienced enough to spot this, was smart enough to talk to her own boss about the symptoms. Once made aware of the nature of the problem, she was able to eliminate it by working on motivation, both individually and during those Thursday meetings. She found, by experimentation, that she had not been passing on enough of the information at her level to the group and that they had become disconnected and worried because of it. Communication was the cure.

## RON, ALONE

Ron is a software person. He is a loner and always has been. He was a little withdrawn and nerdy in high school, and he didn't change surrounded by similar types in engineering school. The job, as far as he's concerned, is completely between him and a terminal. He arrives late, stays very late, and eats whatever's portable.

Donna is Ron's boss. She has been trying to socialize Ron. In group meetings, he attends but never stops scribbling stuff. She makes assignments that require Ron to work with others. Whenever she runs into him, she tries new ideas. She knows that she has a seven-person group, *plus Ron*, and that there has to be a loss of productivity there. Donna is also mildly concerned that Ron's personal development could be improved.

Donna has the right idea. There is no synergy among loners, and there's no group. It's also much more work to motivate and guide people individually. The question is, how best to bring Ron around?

One way is to separate him from the terminal with an assignment that can be done only by communicating face-to-face with others. Such an assignment might be gathering information for planning purposes, establishing a liaison with the marketing or sales people, or making a public presentation. Drawing Ron into group activity by asking his opinions about tasks other than his own is also a viable approach. Effort is justified; loners cost money.

# Chapter 7

# Making Progress

### HOW DO YOU KNOW WHAT'S HAPPENING?

The Emperor, you may recall, couldn't see his new clothes. Neither could the tailors who were supposedly sewing them up. Nobody let on, however, and the situation became more and more ridiculous.

There are many invisible products in high tech. I'm sure that you've seen software efforts that were not only soft but completely intangible — the fabled *vaporware*. Whole factories have been built to run largely imaginary processes (see box). A good fraction of military contracting is for products that do not have to work or will never be tested. Even good products go through stages at which there is almost no way to know how the development process is going.

Even if the product is visible, some of the work being done in your group can be exotic, specialized, and of unknown quality. The difference between you and the low-tech manager is that he understands all his employees' jobs. This is an advantage.

This chapter gives you techniques for measuring progress and performance under conditions when visibility is not good.

### An Invisible Display

One of the largest electronics companies leapt into the flat screen display business with a largely unproven (that is, uncharacterized and unpiloted) design and process. The company made this decision because its research department had a handle on some promising technology and had fattened itself at the Federal teat developing that technology. Without so much as a single fully operational prototype, the company sold itself on the idea and built a factory to produce millions of displays. Despite extensive warnings from equipment vendors and people like me who would have liked to help the company make money, they charged ahead without a real product and without any players from the display industry. At least 50 million dollars later, a handful of displays were produced, and the doors closed forever.

## YOUR TAILORS

Suppose you have the kind of product that can't be evaluated until it's completed. Is it enough to make sure that everyone turns up and puts in time? It may be . . . if your company is in the nondeliverables business. The rest of us have to make something that runs, makes noise, and sells in a competitive market. This means that the eventual worth of your effort depends on things that are beyond your control, such as the following:

- The quality of the original idea
- The state of the market
- What the competitors have

It's insufficient to have a well-functioning product, if nobody wants to buy it.

Also, I've mentioned before that your whole effort can be wiped out by trivial technical decisions or errors. If you can't check everyone's work, down to the tiny details, how can you avoid disaster?

It's a delicate business, no doubt about it. Here I am telling you that just because you're a manager, you must suddenly develop a deep and comprehensive knowledge of the product, the market, and arcane specialties . . . or else. And there you are, thinking, "Not a chance."

### How to Talk to Tailors

- Insist on visible milestones, tests, demonstrations. There should be no doubt when a milestone is reached. Surprisingly, some of your people will think this is silly. This is because the visionaries among us are always living in future time. In addition, we instantly lose interest in a done thing and can't be bothered with writing it up.

- Treat milestones as default objectives. They should be real and deliverable. This will help reduce vaporware.
- Enlist help. Bring in another expert, a consultant, someone with experience. An independent observer may be able to spot an area that is fading from view.

## STEALTH PROJECT

Nondeliverables are objectives that everyone, or nearly everyone, knows are not going to be met. Mostly these are a staple of government work, but there are a few in the commercial world.

A typical nondeliverable is a basic research project. The stated objective is grandiose, like finding the source of the universe *and* curing heart disease. The plan calls for many specialists, sophisticated experiments, and reports — lots of reports. In fact, most of us who work on nondeliverables actually believe that the reports *are* the product. The typical project is so invisible that it might as well be named "Stealth."

♦ Planning for nondeliverables is more complex than you'd guess. You have to come up with a sequence of tasks that arrives, eventually, at the nondeliverable objective. Milestones are obviously either unverifiable or merely invisible.

The paradox of this kind of work is that, done properly, it results in worthwhile contributions to knowledge and might help some future stealth-worker to actually cure something. Therefore, you have to take it seriously, demand good work, and ensure that it becomes available to others. If you're cynical, you're in the wrong job.

## LEAVING BAGGAGE BEHIND

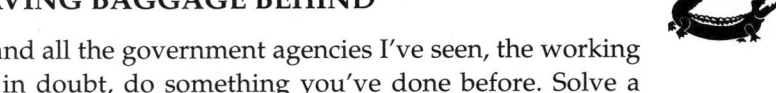

In half the companies and all the government agencies I've seen, the working motto seems to be: When in doubt, do something you've done before. Solve a problem whose answer you already know. Everyone carries some technological baggage around with them. Reinvented wheels, recycled designs, old concepts. In fact, some important and famous people have sung a single song for an entire career, over and over again.

The difference between recycled work and new work is hard to spot. Here are a few tips:

### How to Dump High-Tech Baggage

- New work has errors and dead ends. Any task that goes very smoothly could just be recycled old work.
- Extra features that you don't remember requesting are often vestiges of another project, for which the features were needed.
- Excess complexity results from patching up an old design, or old software, to meet new requirements. I've seen a one-transistor amplifier grow to fill a whole board by this process.
- Sad to say, the older we get, the more likely we are to repeat ourselves. Some of us, however, start very early.
- Some company cultures *encourage* repeating successful work. This is why tiny start-ups can get the jump on the largest companies. When your culture gets that stultified, consider leaving.

## THE NIH SYNDROME

SORRY, MOSES, THEY LOOK LIKE GOOD COMMANDMENTS, BUT THEY DIDN'T COME FROM OUR SHOP

*Not Invented Here.* These three words have caused more duplication of effort, and more waste, than all the innocent errors and wrong turns in the history of technology. The motivation is logical, if wrong-headed. The desire to "own" a concept, a patent, or a product by reworking an extant one, is part of the American Way of Business. Start with Company A's invention — which works. Company B will then go to any lengths to develop, from scratch, a similar but different invention, which may or may not work but which *is* guaranteed to be late. Semiconductor device companies did this dance in design and process for decades, until the associated costs finally beat some sense into them. Kodak engaged Polaroid in an instant film war. The entire American automobile industry provides countless examples of NIH holding back progress. NIH even causes duplication of effort between divisions of a single company, incompatible products, increased development costs, and lower productivity. It's more than a syndrome, it's a whole disease.

Mostly, it's the speed of the high-tech product cycle that makes NIH such a bad influence. Your company must adopt important advances in the field, regardless of origin. At the bench level, where NIH also exists, you need to take action.

---

**The NIH Cure**

- Those of us who don't read the serious literature in our own specialties are likely to have contracted NIH. One indication is seeing something touted in the house organ (company newsletter) or company ads that somehow never appears in refereed journals. If the idea is such a big deal, how come nobody else thinks so? Reading will set you free.

- NIH infests government procurement. Many of us make systems and devices to special specifications that don't really affect the function. The result is continuous production of obsolete technology. There are some signs of improvement. Do your bit for your country. Opt for commercially available product.

- NIH costs more. The higher the tech, the more it costs. You can become a company hero by opting for buying or licensing technology instead of reinventing it.

- Professional pride, and the desire to put a personal stamp on the work, breeds bench-level NIH. The engineer who gets pleasure from individualizing a product, the researcher who duplicates and slightly changes an experiment — both are infected. As a manager, you're in a position to satisfy their need for recognition some other way . . . rewarding for real innovation, for example.

---

## RETROENGINEERING

*Reverse engineering, creative copying, outright theft.* All of these are names for disregarding the legal rights of other innovators. Some firms actually have, as their only business, the careful dissection of existing products so that plans can be sold to other companies. Other companies do it all in-house.

Roger is a retroengineer. His specialty was failure analysis — that is, he took stuff apart to see what went wrong. Now he takes competitors' products apart to copy them. His tools are gleaming analytical machines and computers. He doesn't have to dig through the competitor's Dumpster or meet industrial spies in dark alleys. His company obviously approves of what he does.

Yet, Roger is losing enthusiasm. His significant other thinks that he's just stealing, and he doesn't tell his friends exactly what the job is. He has been wondering lately whether inventing things is worthwhile if someone like him can get paid to rip off the idea. Ethics and business are getting all mixed up in Roger's mind. Where should he go for help? The company legal department? HR? A priest? The District Attorney?

As development costs rise and product cycles get shorter, more and more theft goes on. A company, while seeming to exemplify the worst of NIH, is happy to steal designs, proprietary ideas, patents, and of course, employees. As segments of industry decay and become unprofitable, more litigation and fighting takes place over these thefts. In fact, one of Roger's colleagues in the art cut a contract with his employer that makes the employer responsible for his legal defense, if necessary. Roger has to make a personal decision. Either he's doing right, or he's doing wrong.

## NOT TO HURT, NOT TO KILL

Jerry supervises ten chemists in the central research lab of a large company. For the past several years, the work has involved handling a very toxic and carcinogenic material, together with use of hundreds of gallons of dangerous solvents. The lab is a little old and doesn't have adequate fume hoods. Both of Jerry's predecessors have actually predeceased through cancer, and some of the chemists are sick.

The head of facilities will not cut loose any money for better ventilation or even agree to measure airflow. The best that he can come up with is instruction to use only one hood at a time, closing the rest to preserve airflow in the open one. The lab, although large — about 1,000 employees — does not have a safety officer and does not maintain relevant equipment. If OSHA or state compliance officers were allowed on the premises and told what was being handled, they'd shut the whole place down, probably forever.

Jerry understands that he can be held personally responsible, as a manager, for negligence leading to employee injury. Beyond that, he doesn't like the idea of endangering his people because someone won't put in larger fans. What does he do?

Safety is a delicate issue in high tech. The uninformed, and that's most of us, have some vague idea that government agencies are keeping our workplaces safe and that scientists and engineers understand the hazards they create. This isn't at all true. Government enforcement is spotty, and it's easy for most large employers to avoid inspection by a variety of ruses. Your colleagues have no easy way to evaluate the messes they make, and your company can't afford a polymath biomedical wizard to help them.

The unique problem in high-tech is that the hazards are always new and variable. Chemical exposure keeps changing. New materials are created, many of which have never been measured for toxicity. We deal with every kind of radiation, from every sort of source. We also have lots of experimental apparatus, with no safety-related controls. On top of all this, our employers don't want to know about it and try hard to eliminate permanent records.

So poor Jerry, who is trying to be conscientious, is faced with becoming a hated whistle-blower if he tries to get help with a simple problem. If he doesn't, he's sure that he's hurting people who trust him, and he knows that he can go to court as the villain. The worst irony is that, although Jerry can, with his own hands, plumb and wire a roomful of lethal apparatus with no one looking over his shoulder, he can't go up on the roof and change a fan. He feels immobilized. What should he do?

In the actual case, Jerry died young, of cancer. No changes were made to the laboratory, and for all I know, it's still that way. On balance, he and his group would have been far better off making a fuss, at any political or career expense, than continuing to be poisoned.

## ♦ HOW TO TELL A TOUGH JOB FROM AN EASY ONE

You want to load each member of your group fairly. That is, everyone should have enough work, but not a crushing amount. When we do such mysterious jobs, it becomes hard to gauge the size of a task. Tasks tend to look similar when they are just little boxes or lines on your plan. In fact, you may be driving one engineer into burn-out, while another one is rumbling along at idle.

When you have done the task yourself, you can easily gauge the size of the task, but this is no longer a reasonable expectation. The days are gone when any manager could do all of the subordinates' jobs. You may not have much of a feel for most of these tasks. If you have done this kind of work before, you may have done it so long ago that the experience is not relevant.

### Evaluating Tasks

- Tough is in the eye of the beholder. People who are bright or who respond to challenge will eat up difficult assignments. On the other hand, your marginal employee will break if he has to give up a weekend for the cause. Know your people.

- Tough is the way we like it. An aggressive start-up or a Tiger Team may work flat out all the time. *Exactly the same people in a more leisurely setting will balk at overtime.* The company culture, the group norm, and your influence all control the perception of what's difficult.

- A moderate task can suddenly turn mean if the key individual leaves, resources get cut back, or the schedule has to be accelerated. Learn to compensate by shifting the load.

- Rely on your plan and its frequent navigational fixes. If a task is lagging or is error-prone, something's wrong. Maybe you underestimated its size or overestimated someone's capabilities. Either way, frequent and valid measurements of progress will help identify problems early enough to fix them.

## IS YOUR GROUP GOOD, BETTER, BEST?

Comparing your group to other groups is both necessary and useful. It's necessary when you're competing for company resources, raises, and attention. It's useful when you're planning for improvement. Some companies allot resources according to an internal grading system. Under this system, each group is evaluated in a performance review, relative to other groups. The evaluation is done by the next higher level of managers. The highest grades result in rewards, and the lowest result in abolition, consolidation, or . . . change of manager. The same system is used to grade divisions and other larger components of the company.

To be blunt, performance equals profit. We're used to seeing profit as the main virtue of whole companies, and the reward of profit is continued investment. We are not used to applying the word to engineering a component for a goose's bridle. Not our department! All we do here is (research, development, engineering . . .). We don't sell the eventual product.

Wrong. Your bosses are going to come up with a dollar figure that describes exactly how much good your group did for the company bottom line. You and your group will be heroes or bums on the basis of this figure. If this grade is so important, wouldn't it be a good idea to know how to better it?

### ♦ How to Be the Best Group

- Quantify the value of your product. Connect dollars earned, or saved, with objectives completed. Many times this will seem to be a crass or doubtful task, but try anyway. Think of the good résumés you've seen. The achievement portions always mention money and are almost always a little inventive on this score. You have to be equally bold in your reports.

- Find out what numbers are used in evaluations. Often the ratio of *this* to *that*, expense to employees, space to head count, etc. You may have a small group operating a large, expensive piece of equipment, and you bill its costs only to the company. A research particle accelerator is an example. You will, in any expense-based evaluation, look bad. Getting someone else to pay the freight makes you look better. Finding a 100 percent sponsor makes you look best.

- Give your company a public boost. This could be anything a public relations wizard can use, from winning a science or trade association award to producing a benefit for health, environment, or political relations. Don't exclude yourself from this. Even the part of a goose's bridle mentioned earlier might have some eventual use in reducing energy use in goose-drawn vehicles, or it might lessen the on-the-job injury rate to fowl. This PR stuff may be a little far-fetched, but it raises the company's stock price, so it has to raise yours.

- Keep off other people's toes. Neophyte managers often think that advancement is based on making other groups look bad, raiding their employees, "improving" their work, and producing a great deal of dust and commotion. Sure, it's satisfying to outrun another group and beat them at their own game, but *their boss grades you!* I'm not saying that it's always true, but the combative group, or manager, is often seen as a detriment to order and control.

- Produce.

## GRADING YOUR PEOPLE

Just as your group is graded, you will have to grade your own people. These grades will be debated by managers at your own level, because they impact resources. They will also be used in giving the staff motivation, direction, and reward. This is not a simple report card activity.

One company insists on a bell-shaped grading curve for all performance evaluations. This means that, even in the brightest, strongest, hardest working group, someone has to be elected the star . . . and someone becomes the goat. It also means that a bunch of noncontributors in an overhead area has to show the same distribution of worth as your group of selected high performers. It doesn't seem fair that the best drone, in the drone department, gets the same raise as the creator of next year's great product. Fair and policy are two different things!

Grades turn up as part of a *performance review*. The review may be an annual process, or it may occur more often. It is dreaded by the weak, anticipated by the strong, and ignored by the mediocre. It may involve self-grading, plans for improvement, a variety of punishments and rewards. A manager can look at reviews as a waste of his or her time or as an important component in controlling the group. Ten minutes per review or an evening of careful thought.

PERFORMANCE REVIEW ON A CURVE

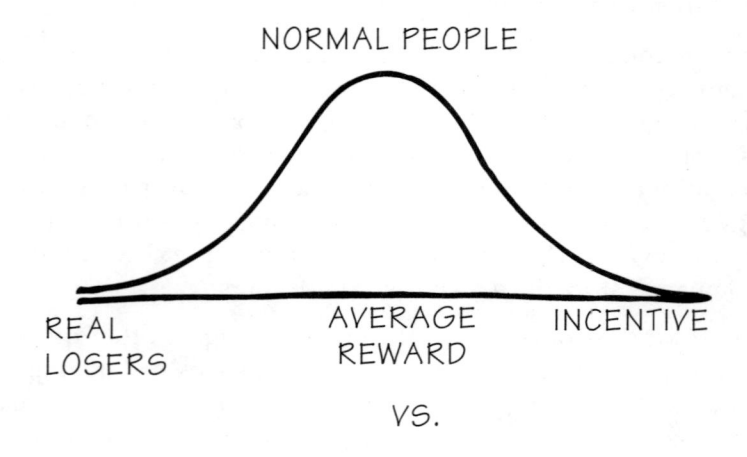

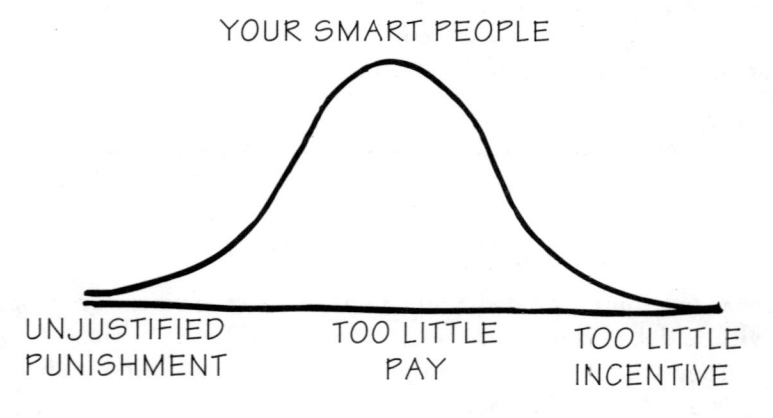

NORMAL PEOPLE

REAL LOSERS           AVERAGE REWARD           INCENTIVE

VS.

YOUR SMART PEOPLE

UNJUSTIFIED PUNISHMENT     TOO LITTLE PAY     TOO LITTLE INCENTIVE

The high-tech alligator is diversity of motivation. Some of your people have unusual ways of getting motivated. A raise — in fact a salary — may be inconsequential to a few of us. Acceptance by others may be a meaningless goal, and you can forget about the sections on neatness, promptness, and physical appearance. So, if you look at most packaged review forms, you should see that there are some categories that do not apply.

Consequently, your staff looks at performance reviews with a mixture of perspectives. To some, the annual review is so important that, when it's late, worry sets in. To others, it's a joke, because it doesn't represent anything which makes the job better or easier.

Assuming that you already know how your company wants you to conduct reviews, the next sections present some thoughts to mull over.

### Get Yourself Calibrated

Is the project going well? Are your bosses happy? Customers? How about you? Is the group's overall performance meeting your expectations? Ask these questions frequently when evaluating individuals. If you don't, you'll drift off toward pessimism or optimism.

### Everyone Isn't Mediocre

A tendency of the careless is to grade everybody as average. It's also a tendency of the newly promoted. There's just something about giving your employee an "excellent" when you yourself have been advanced into ranks where you may be considered as average and inexperienced as well. Give praise when you can and criticize where it's appropriate.

### Recognize the Halo Effect

This is your carrying over a basic opinion into areas where it doesn't apply. For instance, your best engineer might be a poor communicator or abrasive. Because she produces, you're tempted to grade her high in all areas. This is wrong, because it doesn't do her any favors and possibly does harm. She could, for example, be promoted to management.

### Worry About Categories That Do Not Apply

The company culture may encourage sloppy work and dress. Start-ups will, just to project defiance of convention. Dinosaur companies won't promote anyone who looks or is sloppy. Why do some categories seem important in your surroundings and others less important? If you excoriate a scientist for working irregular hours, what does this mean to the company or to the employee? Don't be glib about this. You are, after all, in business, and you don't get to invent *all* the rules.

### Evaluate Often

A good manager doesn't save up a year's praise or criticism for the performance review! That's just as poor as evaluating progress on the project once a year. Communication is a constant process. All of us work better when we know where we stand.

### Emphasize Diverse

Some of your people don't care about acceptable performance in nontechnical areas, such as working well with others. They really don't expect, or think about, promotion. You have a hard time convincing them of the validity of the review or its relevance. Don't take the easy route and write these people off. Beat on them a little. After a few years and a burn-out or two, even the nerdiest among us may want to do something else for a living — such as management, marketing, or sales. Your insistence on developing seemingly irrelevant skills may be appreciated later on.

### Stress Long-term Benefits

"Do extremely well at review time, and we'll let you stay around." This is the reward at heads-up companies. Some very challenging organizations may have more than 50-percent annual turnover of professional staff. Career growth, becoming well-rounded, learning a variety of skills . . . these are not important. Survival is. If you're lucky enough to work in this environment, try to deemphasize it at reviews. Always hold out the promise of long-term benefits.

### Handle "Goats" Carefully

Your company grades on a curve, and you have to elect a goat from your group. If the goat is a good and valued employee, you're responsible for not ruining his or her career. How you do this varies with the situation, but one workable technique is to arrange an internal transfer to another group in which your goat will be guaranteed better marks. To ask someone to transfer out for such artificial reasons is heartbreaking, so you have to explain the reasoning carefully.

### Don't Use the Review to Get Rid of a Staff Member

Sometimes you *do* want to get rid of a staff member. The performance review is used unfairly to assemble the required three strikes, after which the employee can be ejected. This is wrong for several reasons, not the least of which is that the process may be too slow. The wasteful legal situation (employees actually sue, and win, for "wrongful discharge") also pushes us in the direction of this kind of silly documentation. Everyone comes out better if inappropriate staff can be convinced to move, before accumulating a negative, and permanent, file.

### Don't Forget the Law

Because you're new to giving reviews, spend a little time with the personnel department, or the administrative manual, to make sure that you know what types of comments are actionable or against company policy. You will be astounded by what you find out.

### Prepare Carefully

Make a list of specific topics to be discussed in the review, possibly separate from the form you may have to use. Use these topics to guide the discussion. You want to cover the following bases: what was done well, what wasn't, areas for improvement, excuses, communication skills, potential for expanded responsibility, salary, and other rewards.

### Leave the Shop If Possible

How and where you deliver a review is usually established by your company rules. You may be required to summarize a review on a form, which is then signed by the employee, and you may be asked to perform the ritual in conjunction, with the issuance of annual raises. Beyond these constraints, you can make as much or as little of the process as you want. The time might range from five minutes to several hours per worker. Physically getting away from the workplace is ideal, and company travel gives both of you an uninterruptible block of otherwise nonproductive time. If you're in a California-type-of-company, the running track might be the proper venue. In any event, if you want to get someone to take a long view of job performance, or of life, distance from the shop is a good idea.

## SALARIES AND OTHER REWARDS

It's no new phenomenon that engineers and scientists get paid badly, when compared with others of comparable skill and qualification levels. The only new development is that there are many more of these people than ever before and that they're just starting to become aware of low pay. There are even the beginnings of labor unions among those who were raised to despise them.

♦ You can't reasonably expect to reward a low-level engineer with a big raise, a substantial bonus, a company car, or a country club membership. You can't even promise a 40-hour work week or reliable and continuing employment. Just what kinds of motivation can you provide?

### Dual Ladder

Some version of a scheme to give technical staff a "parallel career path" exists at companies that worry about such things. The idea is that you award titles and allow the establishment of a second organizational chart for technical people only.

A typical title might be "company fellow" or "internal consultant." These schemes rarely work, because there is no authority connected with the title and little money. Furthermore, these titles are non-negotiable in the sense that they don't mean anything at other companies. I have much more to say on this subject, but for the time being, the advice is: Be wary of dual-ladder promotions as rewards.

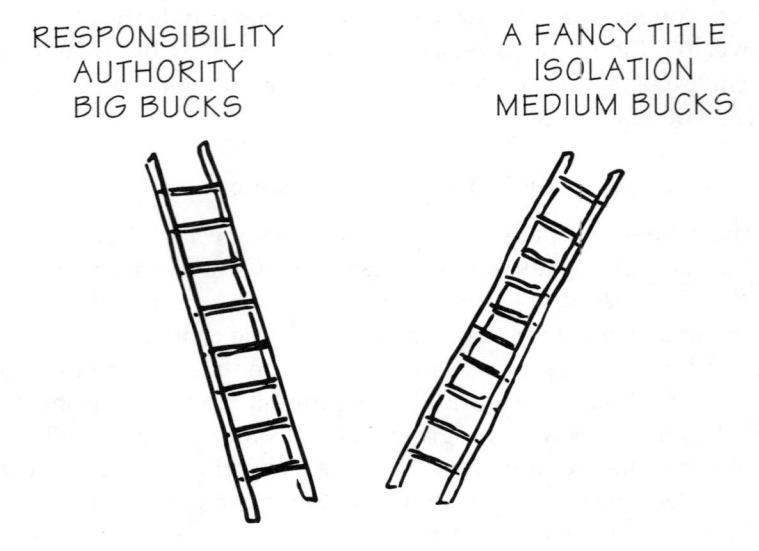

```
RESPONSIBILITY          A FANCY TITLE
  AUTHORITY               ISOLATION
  BIG BUCKS              MEDIUM BUCKS
```

THE DUAL LADDER

### Piece of the Action

Smaller and start-up companies don't seem to mind handing out shares of stock, bonuses, informal profit sharing, and other goodies. Reward can be directly connected to achievement: finish a project, get a contract, ship a machine — get cash. Larger companies aren't as effective at doing this. Their stock purchase plans, profit sharing, and performance bonuses are nearly always a bad bet, or at least not strong motivators. What does an engineer get? Usually a chance to buy company stock a few percent below market price, one U.S. dollar for a patent, and little else. This doesn't encourage good people to excel. Do what you can to make a direct connection between doing a great job and succeeding when the company succeeds.

### Salary or Hourly?

Now that we've noticed that there's a lot of unpaid overtime in our business and that many of us earn pitiful hourly wages because of it, we are unhappy. As a manager, you have to ask your people to put in long hours, for no extra pay, but you also have to expect that this will cause problems. You have a few options: First, nothing prohibits professionals from earning hourly pay. The only virtue of a fixed salary is that it exempts us from protective labor laws (yes, that's why you're an "exempt employee")! See if your company is progressive enough to consider hourly pay. Second, there are always ways to cut contracts with employees for so-called unrelated tasks, done after hours.

You should always try to figure out creative ways to keep those who are working very hard from losing pay rate for doing it.

### A Keg of Beer Is Worth a Thousand Bucks

Having a party to celebrate an achievement at work is a solid way to reward your group. For a modest outlay (which doesn't even have to come from your own pocket), you generate group identity, mark progress, and motivate. It's that basic tribal stuff again.

### Perks for You and Me

Travel to conferences, lab toys, computer facilities, and information tools are on the list. Athletic facilities (as modest as a ping-pong table, as grandiose as a gym) and free snacks (anything with sugar and/or caffeine) also make the list. A salesperson may not be overjoyed at being allowed another long-distance trip, but a bench-bound engineer will jump at it. Toys are always appreciated, especially if they're not totally necessary for the work. Information tools, such as connection to databases and libraries, are considered attractive. Even something seemingly unrewarding, such as being allowed to drive the company truck on a local errand, can represent a break in the routine, hence a reward. Another perk, but one that can be dangerous, is allowing your people to use company facilities for personal matters, like repairing something or using a computer. When this gets out of hand, you wind up with television sets in the lab, viruses on your network, connecting rods in the shop, a dog chewing on the cables, and other problems.

### Extreme Cases

I've come across several start-ups in which intelligent and otherwise sensible people worked up to a year *without pay* after the funding ran out. They were either self-deluded or exploited by their managers into thinking that personal sacrifice would turn the corner for the company and that the subsequent rewards would compensate them. In other companies, marketable engineers have worked for ridiculously low salaries for similar reasons. Although this kind of abuse is not peculiar to high tech, it is something to watch out for. Don't take advantage of people merely because they'll let you. It will always backfire.

At the other end of the spectrum, a company can give a tremendous chunk of cash, or stock, to an employee or prospective employee for doing something unethical or illegal, such as stealing a competitor's product. I know of a few examples, and it's not restricted to upper management. The nature of high tech means that almost anyone can do it, and almost anyone does. Whether this practice is right or wrong is immaterial; the point I'd like to make is that it offends and unmotivates other employees.

# Chapter 8

# Paying Your Way — Every Day

### DOING THE MANAGER THING

What you are is a manager of technical people. What you used to be is a technical person. What you *are not* is a full-time technical person.

To pay your way as a manager you have to control and direct your staff with *leverage*. You have to get more work done than would be possible without you. You have to solve problems — social, administrative, and technical — that are beyond anyone else's skills or responsibilities. In order to accomplish objectives, you have to construct a plan and measure progress against it. While you are at it, you also have to fight fires, keep your bosses happy, and find time to grow your own career.

As you've seen, it's difficult to keep from returning to purely technical work out of habit and for security. It's also difficult to pick up skills in subjects (such as finance or social interaction) that may not interest you.

You are evaluated, however, on how well you do all these things and especially on how much money your skills are worth to the company. You have to demonstrably pay your way, every day.

### MEASURE YOURSELF

Those who do a really superior job in any field, from the violin soloist to the gymnast, have one common characteristic: They all test their own performance constantly and critically. Without being graded by someone else, outstanding performers are able to identify what they did wrong, and they have the ability to work on weak skills. They measure themselves against a very high standard, and if they meet the standard, they create a higher one.

To get to be a good manager, you have to measure how well you're doing and continuously try to improve. In other jobs, other times, an effective system existed for helping you along. Training, however, is not a likely prospect these days. You have to do it alone or with minimal help.

Some of us feel lost when we're no longer making direct technical contributions. When we don't keep daily records, design, apply for patents, write papers, or do the other tasks scientists and engineers do, we feel a little uneasy. "Pushing paper" is the most dismal description of the manager's job. We have no idea how to tell if we're accomplishing anything, let alone enough. The following box presents some suggestions for looking at your own work.

---

**How to Measure Your Work**

- Is the project on plan? The contract you have with your company says that you will deliver certain objectives, so the primary measurement is whether or not the project is on track. In the simplest form, if the milestones are happening, the resources are being expended at the projected rate, and nothing significant has happened in the external world to render the project useless, you're doing your job.

- Are you facilitating? Every day, you should identify several issues or problems you have helped someone else handle. Have you released a purchase order, negotiated for the use of some resource, obtained an authorization?

- Are you directing? Have you compared the plan to the actuals and taken some action, redirecting work to bring the project back to its course? Did you need to do this, or would it have happened automatically?

- Are you communicating? Have you caused an idea, a concept, or data to move from one person to another? From your group to another group? From you to your boss?

- Are you increasing synergy in your organization? Have you assigned work to the right people, discovered and remedied an interpersonal conflict, enhanced the sense of mission?

- Have you used leverage on a technical problem? Have you assigned a team to help you solve a problem by doing background and detail work that you then evaluated? Have you motivated anyone to take on a problem that you could solve yourself?

- Have you thought of a way to improve the way you work? A way to save time? To reduce complexity? To increase quality? To communicate more effectively?

---

## REFINE YOUR PLAN

A plan is a living and changeable entity. As you go along, the plan will go through decision points or nodes, be influenced by external forces, and will be altered to suit current conditions. A good plan will have anticipated many of the changes and their effects. A poor plan will crash when the first problem pops up. As you measure progress against the plan, you're also changing the plan, abandoning dead ends, adding or subtracting resources, and so forth. Another of those pesky new manager problems is treating the plan as a sacred and perfect construction. With experience, you learn to evaluate the plan and refine it, as often as every day.

### How to Keep a Plan Optimized

- Look at it. The worst big-boss thundering I've heard so far came from managers presenting plan updates too casually and uncritically. The tendency is to pull out the plan just before you see your boss and update it. Look at it in between, when you have time to be analytical. What makes sense? What's no longer in the picture? What's new, and is it important?

- Close, but no cigar. A research plan may always have major unknowns; a production plan may be measurable to five decimal places. Your plan has a quantitative deadband where nothing is significant. In other words, the ship's helm doesn't have to be corrected every millisecond; every minute or so will be sufficient. Know how significant a deviation from plan is, and don't waste effort correcting insignificant excursions.

- Update the variables that *aren't on your plan*. That is, the external forces that can change everything. Make a list of developments in the literature and at competitors, political and economic changes, financial events, and anything else that can turn your plan into an academic exercise. Some of these you get from the marketing department, some from customers. Others come from the trade and newspapers.

- Clean up and simplify. Take out the contingencies that weren't needed, combine tasks when possible, and increase the precision of your earlier estimates. If you have five similar tasks, and the first three took ten man-days each, but the plan allotted eighteen each, change the plan for the remaining two tasks.

- Respond to innovation. If there was a breakthrough, many small details of the plan have to be changed. In order to follow up, completely new tasks have to be started, others have to be ended. Breakthroughs are an especial joy, but unless the manager figures out how to fit them into the plan, they can't be effectively exploited and may even be abandoned.

- Play spreadsheet. The one great utility of spreadsheets is playing "what if"— that is, trying out different present conditions. Planning programs all have this capability. Throughout the course of a project you can reoptimize resources, add or subtract branches, expand or contract the time scale. All without much work and without having to learn anything about optimization methods.

## FIGHTING FIRES

Let's say, for the sake of argument, that you are one smart cookie. You like nothing better than to be given a complex, knotty problem to solve and little or no time to solve it. You enjoy knowing a lot of hard facts and also like detective work. This does describe you, doesn't it?

You are thus a candidate for corporate fire fighter. You will be called whenever something unexpected and nasty has happened. You will be found, by expedition if necessary, on your backpacking vacation in northern Canada, and you will be dragged back to the plant. You will learn to keep a filled suitcase at home and a weekend's worth of supplies in your car. You will be asked to fix things in areas of expertise you know very little about. You will probably get results.

You will also neglect your management duties, and your people, while you are out fighting fires. Your spouse will leave you, and your children will not recognize your face.

### ♦ How to Fight Fires Without Getting Burned

Follow these guidelines if you want to avoid being burned while you're fighting fires:

- You may feel useful, since you're called on to solve problems, but your neglect of your own project will be remembered longer. Your primary job comes first.

- The managers who call on you for emergency help are frequently not in your direct line of command. This also doesn't help your career. If, however, you can bring some money into your area by charging to another account, you can get some credit.

- The staff who couldn't solve the problem by themselves are often resentful that their boss didn't have enough faith in them and that *you* solved it. Unless you are masterful at diplomacy, you make some enemies by offering help. The incompetent and insecure are most likely to get upset about it. We know that some of these people will rise above you in the organization. Be humble when on someone else's turf.

♦♦ • If you are successful at fighting a few minor fires, you will be assigned to nastier and bigger blazes until you eventually, and inevitably, fail. You will then be discredited. The trick is to weasel out of suicidal assignments. Try to keep your ego in check as well, because it is easy to overestimate your abilities in this kind of work.

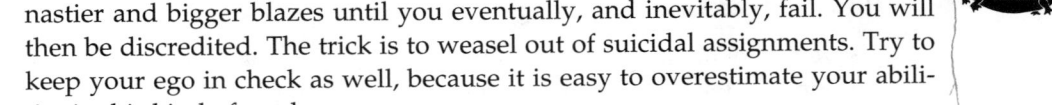

- Never fight fires alone. Very few high-tech problems are simple enough for one person. You need reliable, smart help if for no other reason than checking your work. This is necessary because the normal processes of meetings and group critiques are suspended. You always need other opinions.

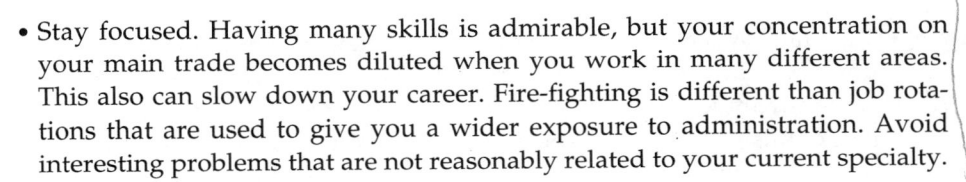

- Stay focused. Having many skills is admirable, but your concentration on your main trade becomes diluted when you work in many different areas. This also can slow down your career. Fire-fighting is different than job rotations that are used to give you a wider exposure to administration. Avoid interesting problems that are not reasonably related to your current specialty.

- After you do it, forget it. The future success or failure of a project that you dug out of a hole shouldn't concern you. After all, you have no continuing say in how that project is run. Think like a tradesman: Do the job, get paid, get out. If you worry about your solution becoming distorted or misused later on, you're asking for trouble. Any continued show of interest or ownership on your part can be received badly. This is a lesson that consultants have to learn in the sandbox.

♦ • You could become a full-time fire-fighter. If enough of the organization above you believes that your best use in the company is as a roving troubleshooter, you may become so identified with it that you lose all opportunity for promotion. If this is okay, go with it. If it isn't, keep this work to a minimum.

## PLEASING YOUR BOSS

You pay your way only when your boss thinks that you do. This does not mean that you have to devote energy to kissing up. It means that your boss believes that you are helping him to do his job, increasing his leverage.

Those of us who have little experience with subordinates miss this concept completely. Our relationship with a boss has been based on taking orders, either willingly or resentfully. We don't think about making the boss's job easier, clearing away debris in his or her path, providing unasked-for help. Part of the re-orientation when we become managers is a growing appreciation of the scope and demands of administrative jobs. You'll start understanding how to help your boss as soon as you need the same kind of help from *your* group.

**How to Make Your Boss Happy**

- Save time. The higher you go in the organization, the more critical time becomes. Helping eliminate unnecessary meetings with your boss by anticipating and acting on all the issues is one good method. Taking on administrative garbage, such as purchasing, shipping, and budgetary form-chasing, helps. Also a good idea is risking a little censure by exceeding your authority when it relieves your boss of unimportant duties. I know companies in which any expense over $250 requires VP signature; this is insane.

- Take hits. When you or your group get sideways with another group, don't hide behind your boss. Try to face the abuse yourself, and try to solve the underlying problem yourself. Sometimes, politics being the sort of business it is, someone may attack your boss by sniping on you or your group. When you sense that this is happening, keep your boss out of it.

- Deliver. None of us wants an excuse, no matter how justified. We want a delivered objective, on time, in budget. There are thousands of reasons for failing to deliver, but none of them should ever be seen or heard on a report.

- Respect decisions. If your boss has signed up to deliver a product, he or she is taking the ultimate responsibility for it. You are being trusted to do some of the work. Second-guessing, negative comments, reservations, and hedging are not appropriate. Your chance to comment on the issue comes before a decision is made, not after. If you really have no faith that the job can be done, get off the job! I can't overemphasize this point.

- The Golden Rule. Treat your manager as you would like to be treated by your staff.

## BITE YOUR TONGUE: WHEN THE BOSS SAYS "LIE"

Al was required to attend an important presentation to a customer on the status of a development project. On the way to the meeting, Al's boss pushed him into an empty room and threatened him severely if he did anything but follow the boss's lead and lie about the product. Al was surprised, because he had not been asked to lie on his routine reports. He was also upset because his boss was actually threatening him with firing, minutes before a presentation. Al is an engineer. The way he sees it, there is no question about the truth when it relates to measurements of volts, ohms, centimeters, etc. He doesn't know how telling a customer that something's working will do anything but backfire later. He also fears that it will backfire on him and his group. His boss will probably claim innocence.

Most of us don't like being ordered to lie, and we're not good at it. Lies, however, are always part of the company culture, or part of some individuals' career plans. Lies are thought to be necessary to maintain a competitive position, to obtain good public relations, and to bring in large orders. Companies lie on performance specifications, in annual reports, on résumés, and all through a reporting chain. You will eventually be asked to lie on a scale that bothers you, and, like Al, you may have only a few minutes to think it over. What did Al do? He was so surprised and upset that he remained mute through the presentation, deferring all questions to his boss. I doubt that he was ever again quite as motivated and committed as before.

## LEVERAGE

Quick, how much leverage do you have? If you transferred away from the project tomorrow, how much less work would get done? Can you prove that your salary, plus overhead, plus 20 percent, can be paid for by enhanced productivity?

When you write a memo, taking one hour of your time and a half hour of a secretary's time, does it result in $750 income to the company? Does the memo result in needless and expensive extra work — if only because it requires replies?

Leverage is bipolar. A good manager pays her own way. A bad manager can lose the company a sizeable multiple of his own salary. The down side is larger than the up side. The engineer who wastes a week designing a superfluous circuit wastes perhaps three times his gross salary. His manager, ordering ill-advised work from a group of five, wastes 18 salaries.

### How to Increase Your Leverage

To increase your leverage, follow these guidelines:

- Do your own job. Keep your paws off your people's jobs. Every time you do work that someone else can handle, you're making that work more expensive.

- Tighten up communications. Memos and meetings are expensive and, by themselves, do no work. Focusing your communications on those most involved and minimizing meetings and archival memos help to get more results with less overhead. Management by walking around is a proven approach to sharpening the focus of your communication.

- Be accessible. Each time a staff member has to wait for you to get off the phone, come back from a meeting, or return from a trip, the staffer is being forced to waste money. Requesting everything in writing may look like a time-saver, because you can look over many issues in a short time. However, it wastes time because there are delays before and after you make those decisions. An effective manager is always accessible.

- ◆ Match authority with responsibility. When you assign a task — that is, a responsibility — also assign the related authority. Only the neophyte manager tries to keep all the authority. The result is that both of you waste time asking and granting approvals for otherwise trivial stuff. If you're willing to trust Elaine with critical technical decisions, you should also trust her to buy library search services or tools.

- ◆ Motivate. If the value of motivation is that it increases individual productivity, motivating your whole group must have a spectacular effect. Each worker you can't motivate also has a disproportionate negative effect. Motivating others is one of those skills that do not come easily to technical folk. Work on it.

- ◆ Delegate. Delegation *is the lever* in leverage. You have to delegate the right tasks to the right people, maintain accountability for the results, and avoid overloading and underloading. Just as you shouldn't do other people's work, you also should avoid delegating tasks that are best done yourself.

## KEEP LEARNING

We all fear technical obsolescence. We know that two years out of school, our skills are looking old. We know that a job that is not on the absolute leading edge is like being stranded on a desert island. You're out of touch pretty fast. The big fear of the new manager is not having the opportunity to stay current in a technical field.

That's the bad news. The good news is that you are now a manager, your important skills are managerial, and it's okay to fall behind in the nitty-gritty technology. Managerial skills are considerably longer lasting.

However, continuous learning is mandatory. You've been promoted to a management job without a formal background in the subject, and nobody is going to send you off to school. You have to scrounge your education from many sources, including books like this one. To keep current and improve your skills, you have to keep reading, keep observing.

### How to Learn About Management

You can learn management skills in many ways:

- Learn by doing. This is always the most effective method. There are many trivial and boring aspects to management that you wouldn't consider learning unless you had to. Now you have to. There are also aspects that you didn't think could be learned, such as leadership. When you do the job right, remember what you've done.

- Jump at company courses. Every year, despite the increasingly inflated figures on training expenditures, lower-level managers have less opportunity for formal coursework. Take any available live or video courses. Even if the theoretical content is laughable — which it often is — you will learn a great deal about how the company perceives its own culture.

- Jump at imported courses. These are one-day to one-week short courses given by outside organizations, such as consultants and trainers. They vary greatly in quality and relevance and are often regurgitated and outdated material. However, you gain perspective about how other organizations function and can spot concepts worth adopting.

- Read. The literature on management is vast. Much of it is written for chief executives, real or imaginary. Much of it runs to case studies that are supposed to contain nuggets of universal wisdom. The pop literature is trendy, full of buzzwords, and is outdated very quickly. There are textbooks, books about small components of management (such as holding meetings), books about quantitative methods, and journal articles on topical studies. Very little of the literature is worth a second reading, so think of it as light diversion, like pot-boilers, mysteries, or romance novels. You can easily digest a typical popular book in an hour or two. It's almost interesting.

- Get a mentor. Among other things, a mentor, who is someone higher-up than you and willing to help you succeed, can teach you the management ropes. The way the word is commonly used, your mentor is not necessarily your current boss. Whatever you learn from a mentor is probably specific to your organization (good) and tailored to your present state of knowledge (also good). Lately, mentors have fallen out of favor, so call your friend something else. How do you find a mentor? Some enlightened companies assign them. Others discourage them. My best advice is to look around precisely the same way you find other good friends or companions: Talk to a lot of people until you find somebody you like who likes you.

- Make it a group activity. You're going to work everything else out by talking with your group. You might as well discuss roles, motivation, ethics, and so forth, as they come up on a day-to-day basis. Benefit from their experience working with other managers, in other groups, other companies.

- Get yourself certified. Credentialed, that is, not committed. Once in a great while, you may be in a position to take a sabbatical (or quit!) and enroll in a management school. At the time of this writing, the combination of a technical degree and an MBA is considered desirable, however distasteful it appears. Of course, this will provoke a surplus, and by the time you read this, the combination may be a detriment. You should be aware that management schools are individualistic and have certain corporate followings. Therefore, a distinct danger exists that you may have the wrong school tie, even when it's from a generally respected place. For most purposes, a few continuing education credits (CEUs) can constitute adequate credentials.

## ADVANCE YOURSELF

Most of the previous techniques for paying your way are concerned with self-improvement and becoming a better manager. All of them involve investment of time and money by you and your employer. If you don't have career growth and advancement, much of this investment will be wasted. So, if your company spends a few thousand bucks allowing you to attend an in-house training course, you are expected eventually to use what you learn.

As you become more experienced, your responsibilities will increase, and your payback to the company will increase. This is another clause in the unwritten contract. Most companies also have a policy of "up or out," especially for managers, and it is difficult to survive for a long time at one level in the organization.

Eventually, you will top out, hit a "glass ceiling," or just get old. Only then will you be expected to stop growing and advancing. Women and minorities, especially, are expected to have stunted career growth in high tech. The concept of age is also discriminatory and arbitrary. Thirty-five might be too old for a new hire in an engineering management path. Fifty-five might be ideal for a VP slot. All nonsense, but all real.

### How to Grow a Career

Here are some suggestions for growing your career:

- Get a reputation. Only known, reliable performers are promoted. You have to deliver your projects consistently, dependably. A single, spectacular success is okay, but it's much better to be known for constant quality ... even if it's not very high.

- ◆ • Show versatility. If you have had a variety of somewhat different assignments and have done well in all of them, you are a candidate for expanded responsibility. In low tech, deliberate job rotations for managers are common. In high tech, the practice is less premeditated, but the virtue of flexibility is universally understood.

- ◆ • Make few enemies. Our jobs and organizations change too rapidly and unpredictably for us to risk making enemies in or out of the company. The competitor or vendor you burn off today might be the employer you can't work for tomorrow.

- ◆ • Develop a sense of money. As you go higher in any company, you will be concerned more with money and less with technology. Decisions you make become conditioned more by potential profit, market conditions, and interest rates and less by elegance or buildability of concepts.

- Belong to the company, not your boss. Many of us have ultimately unsatisfying careers because we decided to ride some superstar's coattails. This works when you've made the right choice. If you have this kind of prescience, you have no reason to work for a living! It's common to pick the wrong coattails and have your leader leave, die, become discredited, or turn on you.

- Work for one company at a time. I know people who think that their jobs are primarily a source of salary and resources with which they will start businesses on the side, pave the way for start-ups, cultivate future raidable staff, and so forth. These people are easily spotted, mostly because almost anyone will rat on them, and they are not rewarded at all.

- Change companies. Nearly all the fast-track characters I know have a personal rule like "no more than two years for one job or one company." This seems wasteful, but it comes from a very real condition of the workplace. Advancement in technical jobs has been limited (because of bad management), and many of us have maintained salary growth or increasing responsibility only by hopping from one company to another. Thus, there's an individual version of "up or out." I don't like it, you don't like it, and it doesn't play in Europe or Japan, but there it is.

## THE HARVARD HULA HOOP

Paul, a VP, just came back from a retreading sabbatical at a well-known management school. Having cost the company a couple of hundred thousand dollars, Paul is ready to make it back in increased efficiency, higher quality, and other organizational virtues. The particular concept he has latched onto is minimizing variances.

Ed runs the manufacturing engineering department. As he sees it, his most pressing problem is that development has handed him a raw, incomplete, and undemonstrated product. Everything on it needs rework, and Ed's not even sure that the whole thing will work. Meanwhile, Paul starts hounding him about variances (which he doesn't bother to define) and starts cutting into all the rework loops. It's beginning to look like Ed will have to hand off a dog to production, just to keep those variances down. What would you do if you were Ed?

Management, at the level that's allowed any opinions at all, tends to attach itself to fads, buzzword programs, and whatever song the academic crowd happens to be singing at the moment. As you probably suspect, fads propagate quickly because managers who have no ideas of their own are compelled to make continuous changes anyway. It's especially noticeable whenever a manager gets a new set of responsibilities. Instead of spending time to understand the task, he or she will just pull a reorganization. This is a phenomenon documented all the way back to ancient Rome.

At the moment, two big fads causing the most waste and headaches for the people who have to respond are Quality and Diversity. Next year, the fad might be "Management by _____." In the worst case, your company might change its primary goal of making money to a faddish objective that doesn't. It happens more than you'd expect. The entire nonfunctional economy of what was the U.S.S.R. was based on topical management theories that don't wash.

♦ If you're caught by some supervisor's fad, you should consider that your ability to deliver results has just been slashed. The fad is, at the least, going to give you more work, without relieving you of any responsibilities. You usually can't fight back, but you can keep the negative effects down if you focus on the long term and do what you have to do to deliver the real goods. Ed has to ride out the silliness.

## AN ALLIGATOR IN CHARGE

Shel is an erratic guy. Most of the time, he's a sociable and competent engineering manager with a good record of accomplishment. Once in a while — full moon or whatever — he turns into a raging beast. He may scream at you, throw things, break furniture, or even fire someone. This behavior has not gone unnoticed, and Shel has remained lower in the organization than his competence and seniority would indicate. His supervisors have given up trying to change him, partly because the company president acts the same way. The problem is that he's your boss.

You tell yourself that you're thick of skin, hard of head, and can weather Shel's outbursts, but you worry that he keeps undercutting you with your troops. You don't like being dressed down in front of your own employees, and you've caught yourself more than once just before passing on the same abuse. What should you do?

♦ The alligator boss is dangerous. He or she eventually will damage your career. You can't go to HR about a supervisor; they might do the wrong thing and hurt you. Similarly, you can't go to your second-level boss — that's disloyalty. As a rule, you will lose in a shoot-out with a supervisor. The only remedy appears to be a powerful mentor, either in your own company or with an influential customer. The mentor has to be able to make the case that the alligator does damage to the company or can't be allowed contact with the important customer. Then, Shel might be reassigned, somewhere far from you.

## LAZY PAUL

Dorothy is a mathematical modeling expert in a tiny high-tech start-up. There are 12 employees, and Paul is the boss. Until now, the company has prospered because of a string of research grants gotten by the chief scientist, who is *not* the boss. Real business is now on the horizon — in fact it's mandated by the contracts — and Paul is too laid-back to go after it. He spends nearly all his time studiously avoiding the rest of the staff. It's rumored that he plays video games on those occasions when he's actually in. Dorothy believes that he's only there because he invested a few bucks at the beginning. One of the founders was his brother-in-law.

Dorothy left a heads-up, medium-size outfit to make some waves in this start-up, and she's getting impatient with Paul. She doesn't even know if he can do the job he's too lazy to start. What should she do?

This is another bad boss situation. Everyone would like to tell their bosses what to do, but high tech is different in that most everybody is results-oriented. This means that many of us are impatient with bosses who are too laid back or who retired on the job. A weak boss directly hampers our own ability to get results and is to be despised.

♦ Dorothy just gave up and moved on. What she could have done instead involves mutiny. She could have taken all the key players — that is, nearly everyone, and formed another company. Replacing Paul would have proved unlikely. In a larger company, perhaps, but not in a small start-up.

# Chapter 9

# Quality — Buzzword or Religion?

### QUALITY QUIDDITY

Quality, whatever it may be, is hounding you. Every time you open a trade magazine, attend a company-wide symposium, or talk to a customer, you hear the word. Your company has recently announced a program to get to six sigma, the halls are infested with bearded consultants, and your man in Japan is sending urgent faxes whining about the difference between Us and Them. You don't know what it all means, or if you do, you're plenty worried. And you should worry, because quality means survival.

The most worrisome aspect of the Quality ruckus is that despite your otherwise excellent college education, you can't recall any coursework on the subject. You did study statistics, and you may even use a little statistical process control theory in your work, but what of the rest of it — the poka-yoke, QFD, robustness, kaizen, kanban, hoshin kanri, CIP, AQL, CWQC — and the gurus — Juran, Crosby, Deming, Feigenbaum, Taguchi? You're sure it's vital, maybe in manufacturing, but what does it mean to you?

I come across lowly wrench-turners at Fortune 5 companies who have been told to make their wrench-turning be six sigma. I come across purchasing agents who are required to ask vendors for similarly sigmoid toilet paper. None of them has the foggiest notion of what to do, which way to jump. Quality has become a sort of Holy Grail — everyone's looking for it, but nobody knows exactly what it looks like.

Meanwhile, the new car you've bought — the product of an intensive quality improvement effort — is fundamentally unrepairable six months off the showroom floor. The billion-dollar space telescope is ground so badly that you could spot the error with a razor blade and flashlight. And nobody in this country knows how to make a competitive VCR, TV, or slab of plywood. Is any of this related to quality?

## ♦ QUALITY, INNOVATION, AND HIGH TECH

Whatever quality is, the one indisputable fact is that it is now part of everyone's vocabulary and on every manager's plan. It isn't a specialist's turf any more. (Remember quality control, quality assurance?) However, because all the formalisms grew out of mass production and military products, how they apply to activities like research and development is still uncertain.

Without getting too bogged down in philosophy, I'd like to suggest a definition of quality that might be useful in our business:

*Quality is any property of product or process that makes long-term money.*

The reason I'm promoting this definition is simple. Profit is the *only* quantitative measure of achievement in common use. Nobody has a number to attach to a Picasso drawing, a Rolls-Royce, or a DRAM that measures some intrinsic aspect of quality. The only thing we can say, for sure, is whether or not the product is profitable.

You may not like this definition. Most of the technical people I've asked have some concept of quality as design elegance, use of durable materials, conformance to specification, repairability, superiority of function. These are all valid ideas, but implementing some of them will cause a company to go broke, so they make poor yardsticks. The entire motivation for enhancing quality in high tech is to increase profit. Making a "better" product is one of the many ways to do it but not even the most important way.

When quality and innovation share the same breath, we have to be very careful to define something we can measure. If we can't measure it, we can't improve it. But what is the measurable product of a research worker? How can a new concept be evaluated? How can you put numbers on a plan? These are all deep and sticky questions.

## WHERE DID THE QUALITY REVOLUTION COME FROM?

There isn't space in this book to go into the history of quality, but it may be useful to know a little about its roots. Before the industrial revolution, quality had many definitions, some subjective and others quantitative. If Napoleon's purchasing agent bought guns, he expected that most of them would fire a bullet most of the time. If he bought paintings, the indicator of quality would be how the boss liked them. Everything, however, was handmade to some extent and had to be individually evaluated.

Then, mass production of interchangeable parts and assemblies became obvious as a good idea. The U.S. Army was first to require this, also for rifles. For the concept to work, parts had to conform to *standards*. Because there is always *variability*, as a result of *wear* of tooling, *material inconsistency*, and *operator technique*, there had to be *tolerances* that specified how close was close enough.

*Inspectors* were the specialists who made the measurements. From this reasonable beginning, motivated by an organized and funded need to kill people efficiently, the United States embarked on a weird and eventually out-of-control adventure, which we are now attempting to set right.

The weirdness is that, since the time of Jefferson, who tried to import the metric system, American industry has fought desperately to avoid standardizing anything. As a result, we still have no mandatory Federal standards for important gauges, threads, temperatures, volumes, horsepower, and the rest. How many wire gauges can you find in your handbook? Even the highest of high-tech industries still has to deal with a horrible mixture of English, metric, SI, and trade-association units, as well as company-proprietary ones.

So, when a company making nuts and bolts had to standardize for quality and uniformity, they had to maintain their own standards. They also had to define what limits were appropriate to their market, and so on. By 1900 we were already 100 years out of step. Even Henry Ford had to own all his suppliers.

In this century, products and systems, such as the telephone system, grew in complexity and outstripped the early concepts of tolerances. Moreover, it became impossible to test every part. Thus, we have Bell Telephone, the Mother of High-Tech Companies, to thank for devising:

- Statistical quality control
- Sampling techniques
- Process control limits

The step-function increase in productivity necessitated by World War II stressed every sector of manufacturing as well as research. In order to get more weapons, more quickly (one of the major themes in quality to date), the government made Bell's practices mandatory and codified what we now call traditional statistical methods. The result was clearly successful, but we slipped up.

Industry generally resisted implementing quality control techniques, except on government products, because it required too much overhead and was too difficult. The government, for its part, went overboard and inflated inspection and documentation procedures until there were more inspectors than workers, testing became downright silly, and productivity declined.

We wound up with two independent and inefficient economies, commercial and governmental, and we have them to this day. The effect has been to slow down development in government systems . . . so much so that most of these products are seriously obsolete before introduction. On the other hand, the commercial market became an uncontrolled hodge-podge of partial solutions and techniques.

By the 1970s, roughly, short-term profit became the one true religion of American business, and shipping bad or untested product became routine and was encouraged. The quality control professionals were shunted to some back room and given a set of ancient gauge blocks to play with, while the companies were trying to make microprocessors, pharmaceuticals, and energy.

The eventual dose of reality came from both Japan and Europe. The Japanese developed their own religion, that of getting market share in new industries. In order to do this, they reasoned that bad product doesn't help, and they tried to avoid shipping it. Moreover, they were rebuilding an economy from utter destruction and were able to embrace new methods and technologies without having antsy stockholders objecting. For that matter, the organizations and the entrenched engineers and managers were gone as well. Part of their program was based on effective industrial-governmental cooperation (JUSE, MITI) and part on encouraging experts from the U.S., such as Juran and Deming. The effect was staggeringly successful.

It never seemed to become obvious to our corporate leaders that we were in the process of getting thrashed. As each industry battle was lost, the failure was rationalized as either being inconsequential or unavoidable. First shipbuilding, then machine tools, then cars, then memory chips — all went away without causing national alarm or any trace of industrial cooperation. In fact, even now, the electronics industry, the largest one in this country, is still pussyfooting around with consortia and tiny government development programs and losing ground by the mile.

The so-called "quality revolution" which has descended on all of us in rhetoric, buzzwords, and strategic programs, is, in fact, a very late response to a sledgehammer stimulus. The intent is to restore our global competitive position. The strategies are educational, scientific, and cultural. There's a great deal of wind and not much substance, so far. Aren't you glad that you have to contribute?

## ♦ GOING AFTER QUALITY

Your entire management job of planning, delegation, and control supports quality as it increases company profit. Everything you do that maximizes the return from resources, saves time, and works toward specified objectives, increases quality.

Quality also comes from appropriate products. The most carefully engineered and otherwise well-thought-out product is low quality if nobody needs it or wants it. If it's true that quality is everyone's job, it's partly your responsibility to know as much as possible about your customer and the market.

Quality in manufacturing is considered to be a matter of low reject or error rate (the various sigma levels). When few enough duds are made, expensive testing can be reduced or eliminated. Quality in design and development is sometimes aiming for wide process latitude, foolproof assembly, serviceability, or mean time between failures (MTBF). Quality in concept is the most subtle of all.

A FINICKY PRODUCT

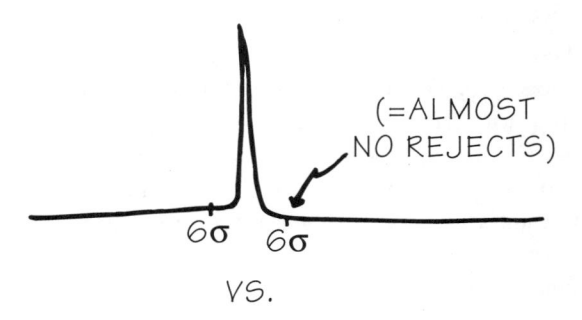

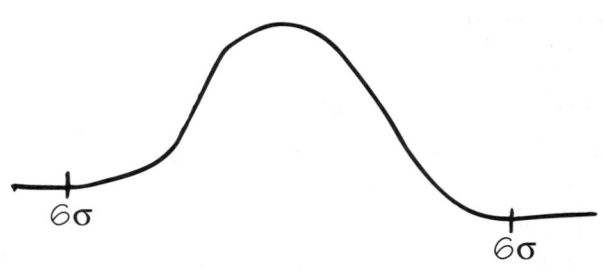

(=ALMOST NO REJECTS)

$6\sigma$ $6\sigma$

VS.

$6\sigma$ $6\sigma$

A WIDE-LATITUDE PRODUCT

WHICH PRODUCT IS HIGHER QUALITY?
WHICH WOULD YOU RATHER BUILD?
WHICH WOULD YOU RATHER USE?

---

Tools which can be applied to quality:

- Design for manufacturability
- Design for testability
- Design for serviceability
- Reliability or failure analysis
- Design simulation
- Prototyping and piloting
- Statistical process control
- Statistical yield analysis
- Testing: life, stress, burn-in

## ♦ USING QUALITY TOOLS

It may be that the best contribution you can make to quality is hiring the most productive staff available. It may be that statistical process control is most appropriate in your job. You will be given the tools you are expected to use, but it makes sense to know which tool goes with which problem.

The first step in doing this is identifying what the problem is. The second step is connecting the problem with a quality-enhancing technique. The third step is applying the tool correctly. These steps are seldom obvious, because:

*A problem and its solution are not obviously related.*

---

**Finding the Right Tool for the Problem**

Keep these ideas in mind when you're choosing a quality tool:

- The problem is always that the customer is not satisfied. The product may be fine but too expensive, not available soon enough, or packaged inconveniently, for example.
- Statistical tools suggest tightening controls; they don't illuminate design or conceptual errors.
- Design-fixing tools can't identify or solve distribution problems.
- You can reduce the defects on the dog to zero; it will remain a dog.
- An otherwise proper quality tool can take too long to apply to the problem.

---

How many times have you seen a problem such as a series of field failures get treated with the statistical ointment? That is, the engineers go back and tighten all the tolerances, or the test people test a larger sample, when in fact the problem was related to a parameter nobody had anticipated?

There's an ancient joke about the new agricultural college graduate who goes off to start a chicken farm. After a few months he writes to his professor: "I bought a thousand chickens, cut them up real small, and plowed them in. Nothing came up. What's wrong?" The professor wrote back, "Send a soil sample."

## HOW TO START A QUALITY-IMPROVEMENT PROGRAM

In a nutshell, treat it like any other planning activity. Set objectives, devise strategies, define tactics, and figure out the best way to measure progress. One great advantage you have is that your company is ready to take on anything labeled "quality."

Another great advantage you have is that, unlike many people in other businesses, your staff understands arithmetic, measurement, and automation, as well as being highly motivated and committed to project success.

Don't forget, however, that the technical solutions you come up with may have to be implemented by production line workers who don't have high school math, may not speak English, and can easily drop the ball.

---

Some objectives:

- High reliability
- High yield
- High customer satisfaction
- Ship no duds

Some strategies:

- Defect reduction
- 100-percent testing
- Faster cycle time, from design to ship
- Robust (solid) design
- Better service
- CIM (computer integrated manufacturing)

Some tactics:

- Training programs
- Using statistical methods
- Methods for motivating and rewarding staff
- Communication with customer/user

---

## QUALITY AS A COMPANY GOAL

The big-time quality consultants generally train and advise the top management level in large companies. It isn't surprising that their advice is that all quality improvement starts from the top and then involves everyone else. This is exactly the way any corporate objective branches out from a single concept to detailed planning.

In the same way, the primary objective can be simple and vague, like "Our company will do everything to six sigma" or "We intend to have total customer satisfaction." As interpreted by various vice presidents, the vague goal becomes more specific and accumulates more data. The people who interact most with customers, for example, may try to determine what makes them satisfied and form strategies to suit. By the time the objective fans out to the bottom level, it has expanded into many tactical details as specific as finding out a customer's favorite color.

If this process works the way it's supposed to, every level of the organization has clear objectives, and all of the work done is complementary. If it is subverted by lack of interest, misunderstanding of the principles, or poor management, then the program will fail.

This is why training and cultural activities are so vital to a quality program. Just to give you a point of reference, companies in this country spend over $200 billion on training, a number that is almost the same as the total spent on elementary, secondary, and higher education. If you make a simple model that includes how many workers there are and how long they work, versus how many students, you come to the conclusion that the average job has the equivalent of two days a week of training, that training is much more expensive than schooling, or that the figures are pure puffery!

The concept that you can take every worker away from his or her job one or two days a week for training doesn't sit well with anyone who can't see a 20- or 40-percent productivity increase as a result. In truth, there isn't really that much training going on, at least in comparison to Japan, where simple assembly-line jobs rate six-month apprenticeships and the elements of quality theory are taught in grade school.

Here, fewer than 20 percent of the graduates of the best engineering schools have had as much as one course in statistics.

What this means is that quality improvement programs often degenerate into sloganeering, posters, and public relations copy. To do it right, a company has to invest a lot of money, most of it in training, and the money tends to get absorbed by other sinks.

### How to Tell Whether Your Company Is Serious About Quality

As a guide to determining how serious your company is about quality, answer these questions:

- What percentage of your group budget is available for training, including both off-site and on-site programs?

- Can your boss give you a one-hour lecture on statistical process control, reliability theory, or a similar topic? Can any of your engineers?

- Have you personally seen the results of quality-related customer surveys? Can you identify the most important aspect of quality according to a user of your product?
- Do you have the opportunity to dissect or reverse-engineer a competitor's product that supposedly represents high quality?
- Can you safely hand-carry a quality-related concept to someone above your immediate supervisor? What's the equivalent of the "suggestion box?" Does it work?
- Most important: Is your company more profitable than every other one in your field?

## QUALITY IS A COOPERATIVE EFFORT

Lack of noncooperation, the NIH syndrome, secrecy, the antitrust laws — all of these have an impact on delivering a quality product or service. Our system of business pits company against company, division against division, group against group. Any information that arises is kept proprietary and secret. Industry consortia are toothless because, if they were effective, they would violate laws designed to protect us from the robber barons of a century ago. Even the Department of Commerce keeps fearfully distant from doing anything constructive to bring competitors together in national efforts. Meanwhile, we whine about how unfair our global competition is, because their governments help identify and pursue commercial objectives.

You can't fix this situation all by yourself, but you can help change your small corner of it.

### ♦ Benefiting by Cooperation

- You know that if your group can't work together, the group will fail. It's the same between groups, between companies, throughout the entire economy. As a manager, part of your job is to make it possible for people to cooperate. You have prima donnas and loners who, despite their other virtues, inhibit cooperation and reduce the effectiveness of the group. Try to change them.
- Make your group more accessible to other groups. If you allow access to your resources, you can more easily use those of other groups. Trying to charge other groups for your expertise or resources may generate nice-looking billings, *but they're all internal and hence imaginary.*

- Increase your technical exchange with other groups, universities, even other companies. At a minimum, you will have to reinvent fewer wheels and chase down fewer dead ends.

- Consider your end customer to be a partner instead of an adversary. The customer wants quality products; you want to supply them. Learn to share data, observations, and wishes.

- Give consortia a chance. You may feel reluctant to lose your best staffer for months and may want to send a less useful person. Bite the bullet if you have the opportunity.

## ♦ QUALITY AND PROFIT

First I tell you that profit is a good measure of quality; now I suggest that, in high-tech, it's an indirect measurement. If you worked in low-tech manufacturing, the effect of a third-decimal-place yield improvement is apparent instantly on the company books. In our business — especially at the front end, in research — the connection between increasing quality and company profit involves a time delay and many development and manufacturing stages.

The other reason the connection isn't straightforward is that accounting systems aren't structured for it. Added training cost looks like loss of profitability, and added specialist personnel look like losses. The accounting department can't quantify higher customer satisfaction, fewer reworks, and long-term effects.

The problem is that financial systems are fundamentally designed to measure and control small excursions in stable organizations. When the company and its industry are always unstable, growing, and changing, perturbations are not important. This problem has been identified as one of the reasons quality programs are difficult to implement.

On the other hand, we know that poor quality typically costs 20 to 40 percent of total cost. You can drop incoming testing when incoming quality increases enough. Process yields can always be improved. Rework can be eliminated. Quality does pay . . . it just doesn't show up in the accounting department.

## MEASURING QUALITY

If you can't get immediate feedback from the bottom line, where do you get it? Your customers, of course. Your customers do not care whether you have zero defects or whether your staff studies Deming; the only significant measurement is whether the product serves their purposes.

Next in importance are all the internal measurements: defect rate (or yield), MTBF, material conformity, standard deviation, and similar measurements. All of these are useful in refining process, improving design, and improving functionality. They help you do a better job, but the real measurement is still in the marketplace.

---

### ♦ Tips on What to Measure

What should you measure? Here are some tips:

- Invent a customer satisfaction index. The data can come directly or through the sales department. Identify a parameter that can be quantified, such as percentage share of your product versus competitive ones in the customer's product or process.

- Make a list of the customer's ten most important priorities. If cost, for example, is more vital than reliability, your improvement efforts should concern cost. Reckless quality programs concentrate on better functionality, often when it isn't compelling to the user. That's how you generate military avionics that are so redundant that the plane uses up most of its lift and acceleration dragging this ballast around.

- Statistical methods represent the only way to establish closed-loop control over anything complex, except for you and me. Each measurement, however, has to be related to the entire process in order to have meaning. For example, averaging one defect in ten thousand successive operations at one process step among one hundred steps is vastly different than averaging the same yield in a one-step process.

- Remember that not all of reality is represented by a normal or Gaussian curve of distribution, and that most data reduction methods have built-in bias. In other words, if some measurement has two humps, like a Bactrian camel, you don't treat it like a one-humped dromedary, or you get spit upon.

- Once in a while, tracking some parameter that appears to have no importance whatsoever pays off with a breakthrough. There's a standing joke about the lab worker who always records the barometric pressure, phase of the moon, and similar seemingly irrelevant information. The most expensive troubleshoot I've been on yielded no improvement for months, until someone spotted a correlation with one of these "why bother" variables.

---

## SIX SYGMA IS NO MISTEAK

Jon is in charge of the final assembly and test area in a factory that makes large-scale capital equipment for the semiconductor industry. In the last few years, the product has become much more sophisticated (that is, complicated) and costly. Instead of making a gradual product evolution, Jon's company has been forced to take large leaps in technology and risk everything on a single leading-edge product. This has worked twice before, resulting in market dominance and good returns. However, this time around, Jon senses that the new product is a loser. It's taking too much time, and a lot of critical fiddling, to get each unit running in spec, and it seems that the problems are intrinsic to the design.

The customers are experiencing difficulty keeping the machines running, and this information has gotten back to corporate headquarters. The response, however, has been the instigation of a quality program designed around tighter component tolerances, more incoming test, and more testing in general. In Jon's area, he has been asked to increase the burn-in time on the fully assembled product from one day to three. Very small positive effects have resulted from the program, but Jon can't help thinking that the design itself was ill-advised and quirky. What should he do?

High-tech products often can't be extrapolated from earlier ones and therefore are riskier throughout. The Hubble Space Telescope mirror, for example, was ground and measured with entirely new equipment and was never checked with old test gear. Anyone with major technical doubts is treated like a traitor and ignored, or worse. Once, I had to decide to "rat" on a microelectronic product that would have violated the laws of physics had it worked. The experience was unrewarding.

♦♦ Jon's problem is basically that he doesn't have the authority, or even the skills, to fix the product, but he's being held responsible for putting it in a crate labeled "working." If he goes to management with his observations, he's probably ruining the current plan by bringing bad news and bucking all the quiescent yes-men who have approved the lousy design. Even worse, he may have to imply that the company's quality fix is off-base. In short, despite the technical nature of the product's problems, Jon may have to tell many people that they're doing their jobs wrong!

I have never encountered anyone who became a hero by blowing the whistle on a barking product. You can't help, even if you're right and the product is saved. You will never be trusted again, which relates to loyalty rather than judgment. There are only two solutions — ignore the problem or jump ship.

# Chapter 10

# Checklists

## CHAPTER 1

### First Day

- Define the group objective.
- Determine whether there is an existing written plan.
- Find out whether I will have to come up with a plan right away.
- Find out what money, equipment, and staff resources are needed.
- Meet everyone in the group.
- Ask opinions on the current state of the project.
- Let everyone know my office is open.
- Mention that there will be some changes.
- Keep all current work going.
- Make it clear that I'm the focus of authority and communication.

### First Week

- Test controls: Send a letter.
- Test controls: Have a group meeting.
- Test controls: Solve a technical problem.
- Test controls: Assign short, clear tasks.
- Test controls: Contact other groups.
- Paperwork: Categorize incoming paper, start files.
- Paperwork: Respond to or throw away, do not accumulate.
- Evaluation: Find out who does good work.
- Evaluation: Decide whether the existing plan is reasonable.
- Evaluation: Decide whether the objective is reasonable.

### Second Week and Beyond

- Refine concept of objective with group, with boss.
- Take first cut at plan: some milestones and dates.
- Measure resources: money, facility, human.
- Spend all available time with my group.
- Find out what planning tools are in use.
- Find out what resource acquisition tools are available and whom to contact.
- Find out what knowledge tools are available: library, databank, network.
- Understand delegation as leverage.
- Measure time spent communicating: meetings, memos, calls.
- Stay ahead of paper blizzard; get help from others.
- Set up routine meeting schedule and ground rules.
- Buy clothes if necessary.
- Make no promises or deals yet.
- Focus on the primary objective; avoid distraction.

## CHAPTER 2

### My High-Tech Group

Learn about my group:
- It depends on continuous innovation.
- It responds quickly to market forces.
- It is sensitive to small technical decisions.
- It can be reorganized quickly.
- It employs specialists with rare skills.
- It needs access to knowledge.
- It is lean in structure.

### Knowing the Market

- Read the trade press.
- Talk to customers.
- Talk to people outside the field.
- Brainstorm with my group.
- Use common sense.

### Controlling Details

- Value details enough.
- Make sure everybody knows where their detail fits objectives.
- Create an environment in which details are critical.

### Shifting Focus Quickly

- Lead the way.
- Never grumble about a change.
- Don't wait for the paperwork.

### Employing Specialists

- Regard specialists as almost irreplaceable.
- Know that they know more than I do.
- Respect their oddities.

### Establishing a Knowledge Supply

- List the most probable sources.
- Assign group members to monitor specific ones.
- Discourage digests and reports; encourage verbal transfer.
- Start a group-specific database; add to company database.

### Being a Facilitating Manager

- Concentrate on making other jobs easier.
- Use my authority to add resources.
- Protect my group from paper.
- Coordinate.
- Communicate.
- Control.
- Network.
- Be flexible.

### Management Style

Find my management style:

- __% autocrat

- __% democrat
- The Guru
- One of the boys
- Angelic
- Bastardly
- My own
- Someone else's

# CHAPTER 3

## Company Culture

- Read the administrative manual.
- Find out what the unwritten rules are.
- Find out what people are discriminated against.
- Find out what the old school tie is.

## The Organizational Chart

Develop an organizational chart that:

- Shows authorities.
- Shows responsibilities.
- Optionally shows resources.
- Optionally shows product flow.
- Optionally shows external links.

## Making the Real Chart

- Find out who actually is responsible.
- Bypass buck passers.
- Show functions: who does what.
- Chart my own group.
- Match responsibility to authority.

## Making an Effective Team

- Remove deadwood.
- Avoid narrow specialists unless necessary.

- Match skills and assigned tasks.
- Advertise for solutions.

### Finding New People

- Look inside the company.
- Go through Human Resources.
- Check professional and trade societies.
- Check universities.
- Get help from consultants.
- Use headhunters.
- Try the state job bank.
- Raid competitors.

### Effective Interviews

- Go to Human Resources; learn applicable law.
- Make interviewee do the talking.
- Expose candidate to live meeting.
- Expose candidate to job environment.
- Remember that questions are forbidden; listening isn't.

### Diversity

- Learn about diversity in the work place:
  - Discrimination keeps good people away.
  - Discrimination costs money, reduces productivity.
  - My business is global.
- Learn what our status is regarding women, minorities, older, younger, and ill.
- Learn new laws.

# CHAPTER 4

### What a Plan Contains

Learn what a plan contains:

- Objectives or goals.
- Strategies or means.
- Measurements of progress.

## Objectives

- Define what I started out to accomplish.
- Make objectives as detailed as possible.
- Allow objectives to change over time.
- Remember that objectives are neither good nor bad out of context.

## Strategy and Tactics

Learn about strategy and tactics:

- Strategy is a choice of means.
- Tactics describe the resources used.
- Strategies change over time.
- Tactics have inertia and are hard to change.

## Milestones

- Use intermediate goals as milestones.
- Make sure milestones are scheduled frequently.
- Use milestones that are unambiguous and that have clear results.
- Use milestones to end or begin tasks.

## Resources

- Know my resources:
  - People
  - Money
  - Equipment/supplies
  - Space
  - Knowledge
- Remember that resources are under my control.

## A Good Plan

Develop a plan that:

- Models reality.
- Can be tested against events.

- Is accurate enough for the purpose.
- Is simple enough to be understood, manipulated.
- Is complete enough to incorporate breakthroughs.
- Relates time and resources.
- States objectives.
- Has a way to measure progress.

### Planning for Innovation

- Include foreseeable or breakthrough elements in my plan.
- Include incubation delays.
- Keep the plan flexible to accommodate breakthroughs.
- Leave room for spotting discovery.
- Include many branch points at unknown results.
- Use multiple smaller plans, if necessary.

## CHAPTER 5

### Speeding Up

- Determine how long I have to do the project.
- Find out whether there are new tools that work more quickly.
- Identify any bottlenecks caused by old tools.
- Know whether we can afford a major error.

### Product Cycle Time

Learn about product cycle time:

- It is composed of many intermediate steps.
- It shrinks continuously.
- It is has loops (experiments).
- It is influenced by external developments.
- It is sped up by competition.
- It is sped up by having a common goal.
- It is sped up by simulations and models.
- It is slowed by reinventing the wheel.

## Manufacturability

To assess manufacturability, ask these questions:

- Can it be made with present tooling?
- Can it be made with high enough yield?
- What will it cost?
- Can we improve the design?

## Bottlenecks

- Avoid featherbedding.
- Work around bottlenecks created by company culture.
- Avoid mismatch of new, fast tools with old ones.
- Remember that bottlenecks are always management's fault — they result from poor planning.

## Dead Ends

- Remember that dead ends are a normal part of experimental work.
- Recognize dead ends early.
- Make sure the decision to abandon is clear.
- Keep staff prepared to shift gears.

## A Responsive Group

Build a group that:

- Is fast.
- Is flexible.
- Is not overly specialized.
- Is not tied down to equipment.
- Is willing to barter and scrounge.
- Has no sacred approach.

# CHAPTER 6
## The Tribe

The tribe:

- Trusts me as leader.

- Interacts synergistically.
- Needs to have a distinct identity.

## My Group

Build a group that:

- Is self-organizing.
- Assigns roles.
- Has group and individual expectations.
- Tries to preserve itself the way it is.
- Has a definite boundary.
- Can develop rituals.

## Factions

Remember that factions:

- Are subgroups.
- Are generally destructive.
- Compete with one another.
- Can be managed.

## The Boundary

- Remember that the boundary is not necessarily identical to the group itself.
- Remember that the boundary is vague in matrix management.
- Identify the boundary from communication patterns.
- Reinforce the boundary by physical surroundings, if needed and possible.
- Organize picnics to reinforce the boundary.
- Emphasize my group's common function in the organization.

## Motivation

Remember these things about motivation:

- Motivation and work output are proportional.
- Motivation can be manipulated.
- People in high tech have diverse motivations.

- Motivation can be negative.
- Motivation is my responsibility.

### Motivators

People in my group might want:

- Money for basic needs.
- Approval by group.
- Satisfying work.
- Participation in company success.
- Opportunity for personal fame.
- To do good for the planet.
- Having nice toys to work with.
- Many other things.

### Delegation

- Remember that delegating is necessary to management.
- Learn how delegation relates to leverage.
- Avoid the lure of the bench.
- Never assign one task to two people.
- Match task with worker.
- Match authority with responsibility.
- Use delegation as a reward or a punishment.
- Keep delegation within the company rules.

### Meetings

- Keep meetings short.
- Reduce notes and minutes.
- Start on time; end on time.
- Make sure each meeting has a clear purpose.
- Avoid show-and-tell — it's pointless.
- Use brainstorming.
- Allow questions after presentations, not during.
- Avoid elaborate materials.

- Don't hand out a duplicate of what's on a screen.

## Communication

- Communicate! — it's constructive, facilitative.
- Use whatever new media are appropriate.
- Avoid jargon.
- Remember that most communication is most significant over lunch.
- Keep communication focused on the manager.

# CHAPTER 7

## Measuring Progress

- Remember that measuring progress is difficult when work is invisible.
- Develop especially good milestones and tests.
- Treat milestones as deliverables, avoid vaporware.
- Get expert help.

## Dump the Baggage

- Learn to spot recycled old work.
- Be suspect of extra features, extra hardware.
- Be suspect of extra complexity.
- Be suspect of my company culture.

## Not Invented Here

- Avoid the NIH syndrome:
  - It causes massive waste.
  - It slows down development.
  - It infests government products.
  - It costs more.
- Prevent NIH by finding other rewards for the staff.

## What's Easy, What's Hard?

- Remember that the difficulty of a project depends on the setting and motivation.
- Provide challenge — we like it.

- Remember that easy tasks can be compressed into hard ones.
- Develop a plan that smoothes out the hardest bumps.

### Measuring My Group

- Measure my group — it's necessary to spring resources, raises.
- Remember that performance equals profits.
- Quantify our value to the company.
- Impact the outside world with a press release.
- Keep off other group's toes.

### Individual Performance Review

- Review myself first.
- Remember that everyone isn't average.
- Don't allow halos.
- Review continuously, not just once a year.
- Emphasize personal development.
- Emphasize long-term rewards.
- Don't use a review as a weapon to fire someone.
- Find out what the state and federal laws are.

### Salaries and Rewards

- Understand these aspects of salaries and rewards:
  - We're all underpaid.
  - Dual ladder.
  - Equity interest.
  - Unpaid overtime versus hourly pay.
- Use group socialization as a reward.
- Try unusual rewards: travel, tools, toys.
- Don't exploit a cheap reward — it's unethical.
- Don't reward illegal actions.

# CHAPTER 8

## Gauging My Own Work

To gauge my own work, I should answer these questions:

- Is the project on plan?
- Do I facilitate my staff's work?
- Am I controlling, or just watching?
- Am I communicating?
- Is my delegation producing leverage?
- Am I finding better ways to do the work?

### Optimizing the Plan

- Look at the plan as if it were alive.
- Correct only significant deviations.
- Check what's happening in the outside world.
- Clean and simplify.
- Respond to breakthroughs.
- Play spreadsheet.

### Fighting Fires

- Put my real job first.
- Remember that I can be resented.
- Keep in mind that fighting one fire can lead to bigger and bigger fires.
- Always get help.
- Get in, do the job, get out.
- Remember that full-time troubleshooting is another profession.

### My Happy Boss

- Save him time.
- Take hits.
- Deliver objectives.
- Respect the decisions.
- Apply the Golden Rule.

### Leverage

- Do only my own job.
- Tighten communications.
- Always be accessible.

- Study motivation.
- Keep authority and responsibility matched.

### Learning More

Keep learning, from all available sources:

- On the job.
- In company courses.
- In imported short courses.
- From the literature.
- From a mentor.
- From my group.
- From a management school.

### A Career

- Deliver work reliably.
- Be versatile.
- Step on few toes.
- Learn about business and money.
- Belong to the company first, boss second.
- Work only one job at a time.
- Jump ship when it makes sense.

# CHAPTER 9

## Quality in High Tech

Remember these things about quality:

- Quality makes money.
- Money can be measured.
- Quality can be measured.

### Sources of Quality

Quality comes from:

- Appropriate products.

- Jobs done well.
- Low reject or defect rate.
- Process latitude.
- Foolproof design and assembly.
- Serviceability.
- Original concept and purpose.

## Quality Tools

- Develop a robust design.
- Design for testing.
- Design for service.
- Perform reliability/failure analysis.
- Use simulation.
- Use prototyping.
- Learn about statistical process control.
- Learn about statistical yield analysis.
- Test the product.

## The Right Tool for the Job

- Always use the customer's objective as my own.
- Remember that statistical tools don't show design flaws.
- Be on the lookout for dogs: A zero-defect dog is still a dog.
- Remember that the right tool can be too expensive to use.
- Know what tools are in use on my project.

## Starting a Quality Improvement Program

- Understand that my company is ready for a quality improvement program.
- Be glad that my staff can understand the math.
- Identify the most appropriate objective:
  - Reliability.
  - Yield.
  - Customer satisfaction.
  - Ship no duds.

- Identify the best strategy:
    - Testing.
    - Defect reduction.
    - Faster cycle time.
    - Robust design.
    - Better service.
- Choose my tactics:
    - Training programs.
    - Statistical methods.
    - Motivations and rewards.

## Quality in the Company

- Understand where it starts — at the top.
- Understand that it expands downward.
- Remember that it requires large training investment.
- Find out whether I have a budget for quality training.
- Find out who can explain the statistical information.
- Find out what my customer expects.

## Cooperation

- Make sure my group works together.
- Make sure my group works with other groups.
- Help my company cooperate with others.
- Help my industry work with the government.
- Treat my customer as a partner.

## Measuring Quality

- Invent a customer satisfaction index.
- Quantify a parameter related to it.
- List the customer's ten highest priorities.
- Remember that reality isn't a normal distribution.
- Track a few wild-card parameters on a hunch.

# Suggested Books and Periodicals

There are two reasons for the immense number of management books. First, management is a universal job, so that anyone from a fast-food shift supervisor to the CEO of IBM falls into the potential readership. This is an attractive concept to publishers. Second, the vast literature mostly consists of untestable theories, common-sense concepts, case studies, and undisguised autobiographies. There isn't the same sort of quality control that we see in more "technical" fields.

The short list below is arranged only loosely by chapter. Many of these books apply to more than one subject, which is why they're not named in the chapters themselves. The books cover a wide spectrum, from lightweight throwaways to dense and difficult texts. The visions of the authors are often contradictory, and they don't always mesh with my opinions, either.

I've tried to mention mostly books recent enough to be found in bookstores and libraries. Like novels, management books are often discarded by libraries after a relatively short time on the shelves, because so many new titles appear each year. Also, some of the subject matter concerns fads and topical case studies. People tend not to be interested in the details of Dudtek's rise to fame once Dudtek has faded from the scene.

Periodicals, especially those specific to your type of business, are far richer in usable and recent information. Not only are these magazines useful, but the editorial staff is usually accessible by phone and willing to talk.

## BOOKS

### Chapter 2

Kelley, R. *The Gold-Collar Worker: Harnessing the Brainpower of the New Work Force*. Reading, MA: Addison-Wesley, 1985. See especially chapters 7 and 8.

Kissler, Gary D. *The Change Riders: Managing the Power of Change*. Reading, MA: Addison-Wesley, 1991. Intended for upper management and dense with details useful for companies in transition.

Naisbitt, John, and P. Aburdene. *Re-inventing the Corporation*. New York: Warner Books, 1985. One vision of where we're all heading.

Waitley, Dennis E., and Robert B. Tucker. *Winning the Innovation Game*. New York: Berkley, 1989. A personal approach to thinking about innovation.

## Chapter 3

Belasco, James A. *Teaching the Elephant to Dance*. New York: Plume, 1990. Case studies in overcoming corporate inertia.

West, Alan. *Innovation Strategy*. Hemel, England: Prentice Hall, 1992. A scholarly, philosophical study of (higher-level) strategic approaches to innovation.

## Chapter 4

Byrd, J., Jr., and L. T. Moore. *Decision Models for Management*. New York: McGraw-Hill, 1982. A quantitative forecasting and planning text.

Grove, Andrew S. *High Output Management*. New York: Random House, 1983. Sound advice from a giant.

Kerzner, H. *Project Management for Executives*. New York: Van Nostrand Reinhold, 1990. A textbook on methods.

Thomsett, Michael C. *The Little Black Book of Project Management*. New York: AMACOM, 1990. A lightweight introduction.

Wheelwright, S. C., and S. Makridakis. *Forecasting Methods for Management*. New York: J. Wiley & Sons, 1985. A textbook of quantitative forecasting methods, best on statistical methods, accuracy, form of models.

## Chapter 5

Bell, C. Gordon, with John McNamara. *High-Tech Ventures: The Guide for Entrepreneurial Success*. Reading, MA: Addison-Wesley, 1991. Although intended for entrepreneurs and investors, the description of the unique high-tech (especially computer) environment is superb. See especially chapters 5–8 on technology and successful products.

Kelly, Al. *How to Make Your Life Easier at Work*. New York: Avon, 1988. Small-scale strategies for coping with paper, people, meetings, and stress; useful for anyone.

Lumsden, George J. *Getting Up to Speed*. New York: AMACOM, 1992. A short list of tips for new managers.

## Chapter 6

Callahan, J. *Communicating: How to Organize Meetings and Presentations*. New York: Franklin Watts, 1984. Many adaptable techniques.

Fletcher, W. *Meetings: How to Manipulate Them and Make Them More Fun*. New York: Wm. Morrow & Co, 1984. The psychology of meetings.

Hampton, David R., et al. *Organizational Behavior and the Practice of Management*. Glenview, IL: Scott, Foresman and Co., 1978. A massive textbook on the social dynamics of management, organizational development, decision-making, and communication; a good reference.

Jeffries, J. R., and J. B. Bates. *The Executive's Guide to Meetings, Conferences, and Audiovisual Presentations*. New York: McGraw-Hill, 1983. Can help you make and use better materials.

McConkey, Dale D. *No-Nonsense Delegation*. New York: AMACOM, 1974.

Truitt, Mark R. *The Supervisor's Handbook*. Shawnee Mission, KS: National Seminars Publications, 1991. Not a handbook as much as a minimal guide to getting people to work together.

Walton, Donald. *Are You Communicating?* New York: McGraw-Hill, 1989.

### Chapter 7

Broadwell, Martin M. *The New Supervisor*. Reading, MA: Addison-Wesley, 1990. General concepts.

Evans, C. George, *Supervising R&D Personnel*, New York: AMA, 1969. Dated, but specific to the first-level manager in research; see especially chapter 9.

Skopec, Eric W. *Communicate for Success: How to Manage, Motivate, and Lead your People*. Reading, MA: Addison-Wesley, 1990. An easy-reading handbook for novice managers, with many examples.

### Chapter 8

Carr, Clay. *New Manager's Survival Manual*. New York: J. Wiley and Sons, 1989. Ultralight tips, case studies, and examples.

Jaffee, C. L., F. Frank, et al. *The Art of Managing: How to Assess and Perfect Your Management Style*. Reading, MA: Addison-Wesley, 1991. A workbook for defining and understanding your own style, similar to many other works in this field.

Madsen, Peter, and Jay Shafritz. *Essentials of Business Ethics*. New York: Meridian, 1990. Essays by people who should know.

Rowan, Roy. *The Intuitive Manager*. Boston, MA: Little, Brown and Co., 1986. The use and growth of intuition.

### Chapter 9

Feigenbaum, Armand V. *Total Quality Control*. 3rd ed. New York: McGraw-Hill, 1991. Fundamentals of statistical methods. A classic.

## PERIODICALS

Most of these periodicals will not be found in average company or small city libraries. The best way to keep up or to find specific articles is by using on-line or CD-ROM services such as CompuServe, InfoTrac, ERIC, and others that may become available as the technology changes. Some indexes will provide abstracts and means for ordering full text. You should, in addition, have a few subscriptions (usually free) to trade magazines in your own area of technology.

*R&D Management*

*Technological Forecasting and Social Change*

*Journal of Business Strategy*

*IEEE Transactions on Engineering Management*

*Management Review*

*Research and Development*

*Research Management*

*Sloan Management Review*

*Industrial Management*

*Harvard Business Review*

*Research-Technology Management*

*Electronic Business*

*Electronic Business – Asia*